my revision notes

AQA GCSE (9–1)

COMBINED SCIENCE TRILOGY

Nick Dixon
Richard Grime
Nick England

HODDER
EDUCATION
AN HACHETTE UK COMPANY

Orders: please contact Bookpoint Ltd, 130 Park Drive, Milton Park, Abingdon, Oxon OX14 4SE. Telephone: (44) 01235 827720. Fax: (44) 01235 400401. Email education@bookpoint.co.uk Lines are open from 9 a.m. to 5 p.m., Monday to Saturday, with a 24-hour message answering service. You can also order through our website: www.hoddereducation.co.uk

ISBN: 978 1 4718 5140 7

© Nick Dixon, Richard Grime, Nick England 2017

First published in 2017 by

Hodder Education,
An Hachette UK Company
Carmelite House
50 Victoria Embankment
London EC4Y 0DZ

www.hoddereducation.co.uk

Impression number   10 9 8 7 6 5 4 3 2

Year   2021 2020

Cover photo © Getty Images/Flickr RF

Typeset in Bembo Std Regular 11/13 by Integra Software Services Pvt. Ltd., Pondicherry, India

Printed in Spain

A catalogue record for this title is available from the British Library.

Get the most from this book

Everyone has to decide his or her own revision strategy, but it is essential to review your work, learn it and test your understanding. These Revision Notes will help you to do that in a planned way, topic by topic. Use this book as the cornerstone of your revision and don't hesitate to write in it – personalise your notes and check your progress by ticking off each section as you revise.

Tick to track your progress ✓

Use the revision planner on pages iv to vii to plan your revision, topic by topic. Tick each box when you have:

● revised and understood a topic

● tested yourself

● practised the exam questions and gone online to check your answers and complete the quick quizzes.

You can also keep track of your revision by ticking off each topic heading in the book. You may find it helpful to add your own notes as you work through each topic.

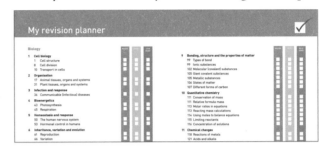

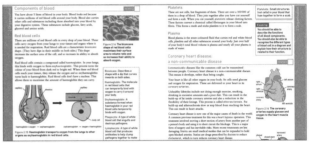

Features to help you succeed

Definitions and key words

Clear, concise definitions of essential key terms are provided where they first appear. Key words from the specification are highlighted in bold throughout the book.

Exam tips

Expert tips are given throughout the book to help you polish your exam technique in order to maximise your chances in the exam.

Now test yourself

These short, knowledge-based questions provide the first step in testing your learning. Answers are online.

Summaries

The summaries provide a quick-check bullet list for each topic.

Exam practice

Practice exam questions are provided for each topic. Use them to consolidate your revision and practise your exam skills.

Online

Go online to check your answers to the Exam practice questions and Now test yourself questions, and try out the extra quick quizzes at **www.hoddereducation.co.uk/ myrevisionnotesdownloads**

Revision activities

These activities will help you to understand each topic in an interactive way.

Typical mistakes

The author identifies the typical mistakes candidates make and explains how you can avoid them.

H Where this symbol appears, the text to the right of it relates to higher tier material.

My revision planner

REVISED TESTED EXAM READY

Exam practice answers, Now test yourself answers and quick quizzes at www.hoddereducation.co.uk/myrevisionnotesdownloads

Countdown to my exams

6–8 weeks to go

- Start by looking at the specification – make sure you know exactly what material you need to revise and the style of the examination. Use the revision planner on pages iv to vii to familiarise yourself with the topics.
- Organise your notes, making sure you have covered everything on the specification. The revision planner will help you to group your notes into topics.
- Work out a realistic revision plan that will allow you time for relaxation. Set aside days and times for all the subjects that you need to study, and stick to your timetable.
- Set yourself sensible targets. Break your revision down into focused sessions of around 40 minutes, divided by breaks. These Revision Notes organise the basic facts into short, memorable sections to make revising easier.

REVISED ☐

2–6 weeks to go

- Read through the relevant sections of this book and refer to the exam tips and key terms. Tick off the topics as you feel confident about them. Highlight those topics you find difficult and look at them again in detail.
- Test your understanding of each topic by working through the 'Now test yourself' questions in the book. Look up the answers online.
- Make a note of any problem areas as you revise, and ask your teacher to go over these in class.
- Look at past papers. They are one of the best ways to revise and practise your exam skills. Write or prepare planned answers to the exam practice questions provided in this book. Check your answers online and try out the extra quick quizzes at **www.hoddereducation.co.uk/ myrevisionnotesdownloads**
- Try out different revision methods. For example, you can make notes using mind maps, spider diagrams or flash cards.
- Track your progress using the revision planner and give yourself a reward when you have achieved your target.

REVISED ☐

One week to go

- Try to fit in at least one more timed practice of an entire past paper and seek feedback from your teacher, comparing your work closely with the mark scheme.
- Check the revision planner to make sure you haven't missed out any topics. Brush up on any areas of difficulty by talking them over with a friend or getting help from your teacher.
- Attend any revision classes put on by your teacher. Remember, he or she is an expert at preparing people for examinations.

REVISED ☐

The day before the examination

- Flick through these Revision Notes for useful reminders, for example the exam tips and key terms.
- Check the time and place of your examination.
- Make sure you have everything you need – extra pens and pencils, tissues, a watch, bottled water.
- Allow some time to relax and have an early night to ensure you are fresh and alert for the examination.

REVISED ☐

1 Cell biology

Cell structure

Eukaryotes and prokaryotes

Eukaryotes

Eukaryotic organisms, or eukaryotes, have cells with a nucleus. Animal and plant cells are eukaryotic. You are a eukaryote, so almost all your cells have a nucleus containing your **DNA** in their middle.

Prokaryotes (bacteria)

Prokaryotic organisms, or prokaryotes, have cells without a nucleus. Bacterial cells are prokaryotic, so bacteria are prokaryotes. All bacteria are single celled and are usually smaller than eukaryotic cells. Bacterial cells do not have a nucleus. Their chromosomal DNA is found within their cytoplasm. Prokaryotic bacterial cells also contain small rings of DNA called plasmids.

> **DNA (deoxyribonucleic acid)**: The genetic information found in all living organisms.
>
> **Ribosomes**: Subcellular structures found in the cytoplasm of cells in which synthesis occurs.
>
> **Respiration**: A chemical reaction that occurs in mitochondria found in the cytoplasm of cells which releases energy from glucose for life processes.

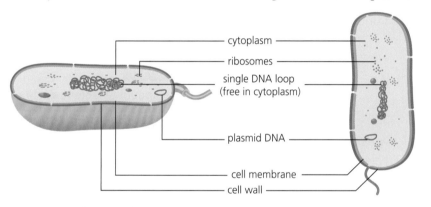

Figure 1.1 A bacterial cell as three- and two-dimensional diagrams.

The components of prokaryotic cells and their functions are shown in Table 1.1.

> **Revision activity**
>
> Draw out this table with only the headings along the top and the first column on the left. Try to fill in the rest of the table from memory to help you to revise.

Table 1.1 The components of bacterial cells and their functions.

Component	Structure and function
Cytoplasm	This fluid is part of the cell inside the cell membrane. It is mainly water and it holds other components such as **ribosomes**. Here most of the chemical reactions in the cell happen (such as the making of proteins in ribosomes).
Cell wall	Like those of plants and fungi, bacterial cells have a cell wall to provide support. However, unlike plant cell walls this is not made of cellulose. The cell membrane is found on the inside surface of the cell wall.
Single DNA loop (DNA not in chromosomes)	DNA in prokaryotes is not arranged in complex chromosomes as in eukaryotic cells. It is not held within a nucleus.
Plasmids	These are small, circular sections of DNA. They provide genetic variation for bacteria.
Cell membrane	This controls what substances go in and out of a cell. It also has internal extensions that have enzymes attached to them. **Respiration** occurs in these enzymes.
Ribosome	Proteins are made by ribosomes, which are present in the cytoplasm.

Animal and plant cells

Generalised (typical) animal cells

A generalised animal cell is seen in Figure 1.2.

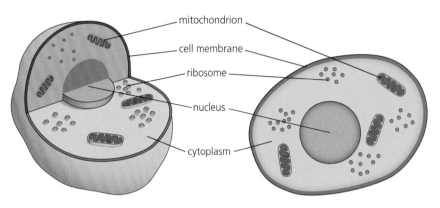

Figure 1.2 A generalised animal cell as three- and two-dimensional diagrams.

Generalised animal cells have components in common with bacterial cells described on the previous page. Both possess a cytoplasm in which most chemical reactions occur, and a cell membrane that controls what enters and exits the cell. The functions of their additional components are found in Table 1.2.

Generalised (typical) plant cells

A generalised plant cell is seen in Figure 1.3.

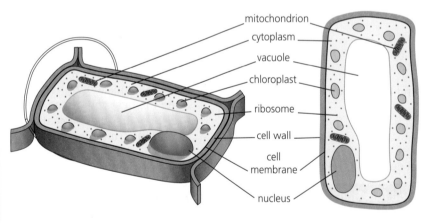

Figure 1.3 A generalised plant cell as three- and two-dimensional diagrams.

Generalised plant cells have all the components previously described for animal cells. Both possess a nucleus, cytoplasm, membrane, mitochondria and ribosomes. Additionally, many plant cells possess a vacuole, which is a store of sugary cell sap, chloroplasts where photosynthesis occurs and a cell wall to provide structure.

Revision activity

Draw out Table 1.2 with only the headings along the top and the first column on the left. Try to fill in the rest of the table from memory to help you to revise.

Exam tip

You should be able to explain how the structures in Table 1.2 are related to their functions.

Typical mistake

The nucleus of a cell is much larger than that of an atom (in chemistry). The nucleus of a cell is made up of thousands of atoms each with their own nucleus.

Organelle: A part of a cell with a specific function.

Mineral ions: Substances that are essential for healthy plant growth, e.g. nitrates and magnesium.

Answers and quick quizzes at **www.hoddereducation.co.uk/myrevisionnotesdownloads**

Table 1.2 The components present in generalised animal and plant cells and their functions.

Component	Present in: Animal cells	Plant cells	Function
Nucleus	Yes	Yes	This contains the DNA or genetic information of an organism arranged into chromosomes.
Cytoplasm	Yes	Yes	This fluid is part of the cell inside the cell membrane, which is mainly made of water. It holds other components like mitochondria and ribosomes.
Cell membrane	Yes	Yes	This controls which substances go in and out of a cell.
Mitochondrion (plural mitochondria)	Yes	Yes	A small cell **organelle** in the cytoplasm in which respiration releases energy from glucose.
Ribosome	Yes	Yes	Proteins are made by ribosomes, which are present in the cytoplasm.
Vacuole	No	Yes	The vacuole is found in the middle of many plant cells and contains cell sap. Dissolved sugars and **mineral ions** are stored here.
Chloroplast	No	Yes	Photosynthesis occurs in chlorophyll found in chloroplasts.
Cell wall	No	Yes	Like bacteria and fungi, plant cells have a wall to provide support. This wall is made from cellulose.

Cell specialisation

You learned about generalised animal and plant cells on the previous page. But complex multicellular organisms like you, and other animals and plants, are not made up of just one type of cell. There are around 200 different types of cell in your body. Each cell type has become specialised to complete a specific function. Cells can develop specific components during this process.

Your cells, and those of many animals, became specialised, or differentiated, before you were born. Unlike animal cells, many plant cells retain their ability to **differentiate** throughout their entire life.

Sperm cell

Sperm cells possess a tail to propel them towards the **ovum** (egg). For their very small size, they have many **mitochondria** to release energy during respiration. Their nucleus contains the DNA from the father, which will make up half of the DNA of the new organism.

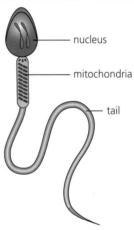

Figure 1.4 The parts of a sperm cell.

Differentiate: To develop into something different.

Ova (singular ovum): Eggs.

Mitochondrion: A small cell organelle, in which respiration occurs, found in the cytoplasm of eukaryotic cells.

Typical mistake

It is important that you can explain why a specialised cell is adapted, not just state its adaptation.

Exam tip

You should be able to explain the importance of cell differentiation.

Nerve cell

Nerve cells (neurones) pass electrical signals around your body to control and coordinate your actions. They possess a long **axon**, along which the electrical signals quickly pass. This is insulated by the **myelin sheath**. Nerve cells possess branching nerve endings that can communicate with surrounding cells.

> **Axon**: The extension of a nerve cell along which electrical impulses travel.
>
> **Myelin sheath**: The insulating cover along an axon that speeds up the electrical signal.

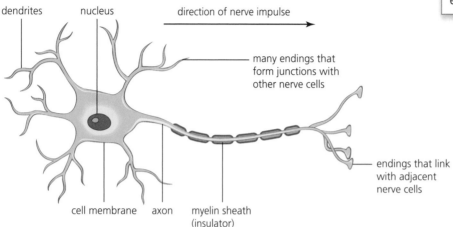

Figure 1.5 **The parts of a nerve cell.**

Muscle cells

Muscle cells contract and relax to move parts of our bodies. This movement can be automatic or involuntary like your heartbeat, or voluntary like moving your fingers to type an email. They possess large numbers of mitochondria to release the energy from glucose in respiration.

Root hair cell

Root hair cells possess a long extension into the soil. This extension massively increases the surface area of the cell, meaning that it can absorb more water and minerals from the soil.

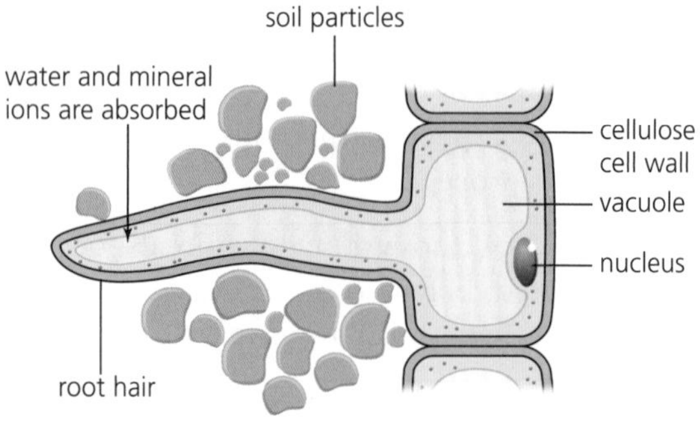

Figure 1.6 **The parts of a root hair cell.**

Xylem cell

Xylem cells form long tubes that run from the roots to the leaves of plants. They carry water to the leaves for photosynthesis. This process is called transpiration. Xylem cells are dead and have eroded-away ends to allow the water to move more easily.

Phloem cell

Phloem cells carry dissolved glucose made during photosynthesis from the leaves to all other parts of the plant. This process is called **translocation**. Phloem cells are alive. They have specialised endings to their cells called sieve plates, which allow water to flow more easily through them. They have companion cells that support the **metabolism** of phloem cells.

> **Phloem**: Living cells that carry sugars made in photosynthesis to all cells of a plant.
>
> **Translocation**: The movements of sugars made in photosynthesis from the leaves of the plant.
>
> **Metabolism**: The sum of all the chemical reactions that happen in an organism.

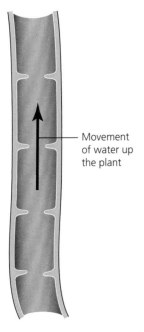

Movement of water up the plant

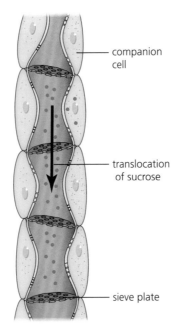

companion cell

translocation of sucrose

sieve plate

Figure 1.7 The parts of a xylem tube.

Figure 1.8 The parts of a phloem tube.

Microscopy

REVISED

Light microscopes

Microscopes allow you to see structures too small to view with your eyes alone. Light microscopes were the first type to be developed, perhaps by Dutch eye glass makers in the 1590s. The design of light microscopes has developed since then but the basic principle of using glass lenses to magnify images is the same. A typical light microscope is shown in Figure 1.9 and the functions of its components are given in Table 1.3.

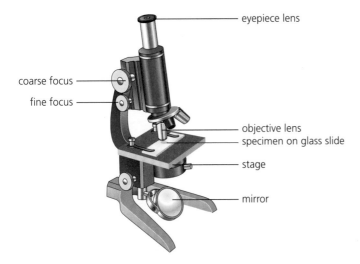

eyepiece lens

coarse focus

fine focus

objective lens

specimen on glass slide

stage

mirror

Figure 1.9 A labelled diagram of a light microscope.

Table 1.3 The functions of the parts of a light microscope.

Part	Function
Eyepiece lens	You look through this lens to see your sample. This is often ×10.
Objective lens	Usually there are three to choose from (often ×5, ×10 and ×25). The smallest will be the easiest to focus, so select this first. When you have focused this lens try a different one with a greater magnification.
Stage	This holds the sample securely, often using two metal clips.
Specimen	This is usually placed in a drop of water or stain on a microscope slide under a very thin glass cover slip.
Mirror	This reflects the light up through the sample, and then the objective and eyepiece lenses into your eyes. In more expensive/advanced microscopes the mirror is replaced by a light source.
Course focus	This quickly and easily moves the stage up and down to focus on the sample.
Fine focus	This sensitively and slowly moves the stage up and down to allow you to make your image very sharp.

The total magnification of a light microscope is calculated by this formula:

total magnification = magnification of eyepiece lens × magnification of objective lens

> **Revision activity**
>
> Draw out this table with only the headings along the top and the first column on the left. Try to fill in the rest of the table from memory to help you to revise.

> **Exam tip**
>
> You should be able to show that you can estimate the scale and size of cells.

Required practical 1

Light microscopy

Aim: To use a light microscope to observe, draw and label a selection of plant and animal cells.

Equipment: Light microscope, slides with samples, drawing equipment.

Method:

1 Place your slide onto the stage of the microscope.
2 Observe using the lowest power objective lens.
3 Use the course and then fine focus dials to focus the image.
4 Change to the next highest objective lens and refocus using the fine focus.
5 Repeat with higher objective lenses if appropriate.
6 Draw scientific images of your observations.
7 Record the magnification that you used on your images.

Electron microscopes

Light microscopes have a maximum magnification of around 1000 times. This means they can be used to see cellular components like mitochondria and ribosomes.

However, in the 1930s German scientists developed the **electron microscope**. This microscope uses beams of **electrons** in place of light to magnify an image. The wavelength of the electrons can be a 100 000 times smaller than light. This fact allows electron microscopes

> **Electron microscope:** A microscope that uses electron beams in place of light to give a higher magnification.
>
> **Electron:** A negatively charged, tiny subatomic particle that is found in shells surrounding the nucleus of the atom.

to take images at higher magnifications. This first electron microscope was an example of a transmission electron microscope, which takes two-dimensional high magnification images.

The design of electron microscopes developed during the 1930s. A second type of microscope called a scanning electron microscope was also developed by German scientists. This takes three-dimensional high magnification images.

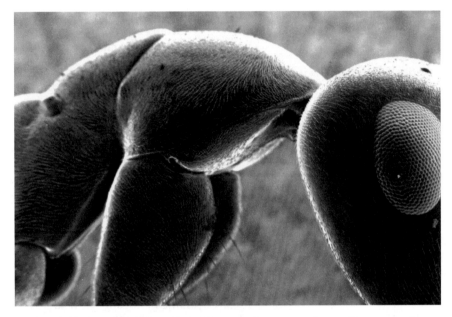

Figure 1.10 A three-dimensional image of an ant taken with a scanning electron microscope.

Magnification calculations

REVISED

The magnification of an image can be calculated by the following formula:

$$\text{magnification} = \frac{\text{size of image}}{\text{size of real object}}$$

Now test yourself

TESTED

1 Define the term DNA.
2 Explain the difference between prokaryotes and eukaryotes.
3 What are plasmids?
4 State the cell components present in plant cells that are not found in animal cells.
5 What process happens in mitochondria?
6 What is the name given to the part of a neurone along which electrical impulses travel?
7 Why do plant roots have root hair cells?
8 Which plant vessels are involved in transpiration?
9 How do we calculate the magnification of a light microscope?

Answers online

Cell division

Chromosomes

Almost all your cells contain a nucleus with one copy of your DNA in. This is your **genome**. Unless you are an identical twin, your genome is unique. It is highly unlikely that anyone has or will ever be born with the same genome as you. Your genome is made up of sections called genes that contain the DNA code to make proteins. Other sections do not make proteins. Currently we are not sure of the purpose of these non-coding regions so we call them 'junk' DNA.

Your genome is made from about 2 metres of DNA. This is too much to be simply arranged loosely in the cytoplasm like prokaryotic bacterial cells (see page 1). It is arranged into 23 pairs of smaller sections called chromosomes. There are 46 in total, but we often say 23 pairs instead. This convention reminds us that chromosomes come in pairs; 23 from our mum in her ovum and 23 from our dad in his sperm. Cells with an entire copy of a genome in them are called **diploid**.

Chromosomes are long thin structures, made from coiled up DNA. They taper in the middle.

Sperm and ova are called sex cells, or **gametes**, and have half the DNA of normal body cells. They are called **haploid**. Two haploid gametes join during fertilisation to make a diploid body cell. You inherit two copies of almost all genes; one from each parent. These copies are called **alleles**.

Humans have 23 pairs of chromosomes, whilst many other animals and plants have different numbers. Lettuces have nine pairs and mosquitos have only have one pair; or one chromosome from each parent.

Genome: One copy of all the DNA found in your diploid body cells.

Diploid: Describes a cell or nucleus of a cell that has a paired set of chromosomes.

Gametes: Sex cells, e.g. sperm, ova, spores and pollen.

Haploid: Describes a cell or nucleus of a gamete that has an unpaired set of chromosomes.

Alleles: Two copies of the same gene, one from your mother and the other from your father.

Mitosis and the cell cycle

Your diploid body cells are continually dying and need replacing. It is unlikely that you have any of the blood cells still alive in your body from a year ago. They have all been replaced at least once since then. Mitosis and the cell cycle explains how this process occurs.

Without mitosis, your body would not have grown from the one diploid cell formed when your father's sperm fertilised your mother's ovum. Additionally, any cuts, burns or other damage to your body would not have healed. Mitosis is essential for growth, development and repair of multicellular organisms.

Mitosis exchanges damaged or old cells with identical replacements. Mistakes in this process could lead to cancer.

During the cell cycle, all chromosomes are copied. This doubles the number of chromosomes from 23 pairs to 46 pairs, or 92 in total. Then the number of cellular components like ribosomes and mitochondria doubles. The cell finally splits into two identical 'daughter' cells, each with an entire copy of the organism's genome.

Steps in mitosis

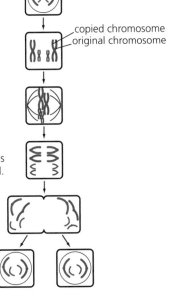

1 Chromosomes make copies of themselves and the nucleus disappears.

copied chromosome
original chromosome

2 Chromosomes line up in the middle of the cell.

3 Original and copied chromosomes move to opposite ends of the cell.

4 Cell divides.

5 New nuclei form in each of the two new cells.

Figure 1.11 The main steps in mitosis for a cell with just two pairs of chromosomes. Note that the two new cells at the bottom of the diagram are identical to each other and also the original 'parent' cell at the top in terms of chromosome number.

> **Exam tip**
>
> You should be able to describe the cell cycle but not the individual stages of mitosis. You should be able to recognise situations in which mitosis is occurring.

Stem cells

Stem cells possess the ability to develop into other types of cell.

Stem cells in mammals

There are two types of stem cell found in you, and other mammals. During fertilisation, your mother's ovum and father's sperm fused. This formed one diploid embryonic stem cell. This divided by mitosis until you were formed. For the first 9 weeks of your life your cells were not specialised. They had not differentiated. They remained embryonic stem cells. These can develop into any of the 200 cell types you possess. These are called **totipotent** cells.

Adult stem cells are the second type of stem cell found in your body. Confusingly, you started to develop adult stem cells much earlier than you became or will become an adult. Adult stem cells are found in specific locations such as the bone marrow and the nose. Here they can only develop into one or two cell types. Adult stem cells in your bone marrow can develop into blood cells, whilst those in your nose can develop into nerve cells. These are called **multipotent** cells.

Other animals like lizards can shed their tails, if caught by them, to avoid being killed by a predator. They are then able to regrow their tail from their stem cells. If one leg of a starfish is severed, it will grow four new ones, and the original starfish will grow one one new leg.

Stem cells and differentiation in plants

Plant stem cells are found in specific locations called **meristems**. These regions are in the tips of shoots and roots. Much of a plant's growth

> **Stem cell**: An undifferentiated cell that can develop into one or more types of specialised cell.
>
> **Totipotent**: Describes a stem cell that can develop into any type of specialised cell.
>
> **Multipotent**: Describes a stem cell that can only develop into several types of specialised cell.

> **Meristem**: An area of a plant in which rapid cell division occurs, normally in the tip of a root or shoot.

occurs in these regions. Unlike your adult stem cells, plant stem cells retain the ability to differentiate throughout their life. This fact means we can take a cutting of a small plant stem and place it in soil. Here stem cells will start to develop into roots and a cloned copy will have been formed.

Stem cell research

Stem cell research uses stem cells to develop future medical treatments that could treat paralysed patients by making new nerve cells to transplant into a damaged spinal cord, or replace injured or non-working organs such as the pancreases of **diabetes** patients. Totipotent embryonic stem cells are more useful in research.

Using a person's own stem cells in medical treatments means their bodies are far less likely to reject them like a transplant from another person. The process of making an embryo with the same genes as the parent for this reason is called therapeutic cloning. However, there is a small possibility of transferring virus infections in this process.

Stem cells research is an **ethical issue**. This means some people disagree with it for religious or moral reasons. Many people donate unused fertilised ova from *in-vitro* **fertilisation** for stem cell research. Controversy surrounds the use of these cells. Are they alive or a life? Because of these issues, tight regulations surround all scientific studies.

> **Exam tip**
>
> You should be able to describe the function of stem cells in embryos, adult animals and plant meristems.

> **Diabetes**: A non-communicable disease that reduces control of blood glucose concentrations.
>
> **Ethical issue**: An idea (issue) some people disagree with for religious or moral reasons.
>
> *In-vitro* **fertilisation (IVF)**: A medical procedure in which ova are fertilised outside of a woman, then placed into her uterus to develop into a baby.

Now test yourself

TESTED

10 Define the term genome.
11 What are chromosomes?
12 State the only two haploid body cells.
13 What are the products of mitosis?
14 What could mistakes in the process of mitosis lead to?
15 What is different about adult and embryonic stem cells?
16 Define the term totipotent.
17 What name is given to an area of a plant in which rapid cell division occurs?
18 In what regions of plant are its meristems?
19 What is the difference between stem cells in plants and animals?

Answers online

Transport in cells

Diffusion

REVISED

Diffusion is the spreading out of particles resulting in their **net** movement from an area of high to lower concentration. This happens naturally and does not require energy, so we call it a passive process. Because diffusion always happens from high to lower concentrations, we say it occurs down a **concentration gradient**.

Particles of gases and liquids can diffuse. Those of solids have fixed positions and so cannot.

> **Diffusion**: The net movement of particles from an area of high concentration to an area of lower concentration.
>
> **Net**: Overall.
>
> **Concentration gradient**: A measurement of how a concentration of a substance changes from one place to another.

Examples of diffusion

Diffusion occurs in many places in your body. You breathe oxygen into the **alveoli** in your lungs. Here oxygen diffuses from a high concentration in your alveoli to a lower concentration in your red blood cells. When these cells absorb oxygen, they turn from a low to high concentration. They then move through your blood vessels to your body's cells. Because these cells have been respiring they have been using oxygen. This means they have a low concentration of oxygen. So, oxygen moves by diffusion from the high concentration in your red blood cells to the lower concentration in your body's cells.

This process is repeated in reverse with carbon dioxide. Carbon dioxide is produced during respiration by your body's cells. It dissolves straight into your **blood plasma**, and is not absorbed by red blood cells, as with oxygen.

Sugars such as glucose are produced by your digestive system when carbohydrates are broken down. These therefore exist at high concentration in your small intestine. So, glucose moves by diffusion from your small intestine through **villi** to the lower glucose concentration in your blood. When your blood absorbs glucose, it turns from a low to high concentration. Blood moves round your body to your body's cells. Because these cells have been respiring they have been using glucose. This means that the glucose now moves by diffusion from the higher concentration in your blood to the lower one in your cells.

Some of your cells make **urea** as a waste product. This diffuses from a higher concentration in the cells to a lower one in the blood. Urea is transported to the kidney where it is **excreted**.

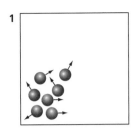
1

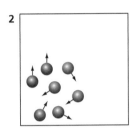
2

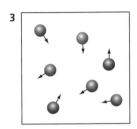
3

Figure 1.12 Molecules in a gas spread out by diffusion.

Alveoli: A tiny air sac found in the lungs through which gases exchange between blood and air.

Blood plasma: The straw-coloured liquid that carries our blood cells and dissolved molecules.

Villi (singular villus): Tiny finger-like projections that increase the surface area of the small intestine.

Urea: A key waste product of protein metabolism in mammals that is excreted in urine.

Excretion: The removal of substances from cells or organisms.

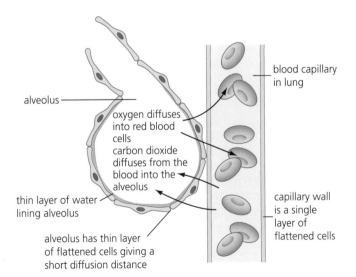

- alveolus
- oxygen diffuses into red blood cells
- carbon dioxide diffuses from the blood into the alveolus
- thin layer of water lining alveolus
- alveolus has thin layer of flattened cells giving a short diffusion distance
- blood capillary in lung
- capillary wall is a single layer of flattened cells

Figure 1.13 Diffusion of gases between an alveolus and a blood capillary in the lung.

Diffusion in other organisms

Your lungs are extremely effective at absorbing oxygen. But not all animals have lungs. Insects have small tubes that run into their bodies, into which gases diffuse. This system is much less sophisticated than your lungs, and so insects cannot complete as much gas exchange as humans. They do not possess efficient exchange surfaces like your lungs or the gills of fish. This means insects are always limited by size. Their maximum size is in part determined by the distance that oxygen can easily diffuse into their cells. In fish gills, water passes in the opposite direction to the blood meaning that maximum diffusion of oxygen into their blood occurs.

Smaller organisms like insects or single-celled organisms have a larger surface area to volume ratio than larger organisms. This means they can survive without specialised exchange surfaces.

Factors that affect diffusion

The factors that affect diffusion are shown in Table 1.4.

Table 1.4 The factors that affect diffusion.

Factor	How
Difference in concentrations	If two concentrations are similar the rate of diffusion will occur slowly. The larger the difference in concentration, the quicker the rate of diffusion.
Temperature	At higher temperatures, all particles have more **kinetic energy**. The higher the temperature, the quicker the rate of diffusion.
Surface area	More diffusion can occur over a large surface area. The larger the surface area, the quicker the rate of diffusion.

Osmosis

REVISED

Osmosis is the spreading out of water particles resulting in the net movement of water from an area of high to lower water concentration across a **partially permeable** membrane. So, osmosis is the diffusion of water across a membrane.

This happens naturally and does not require energy, so we call it a passive process. Because osmosis always happens from a high to a lower concentration, we say it occurs down a concentration gradient.

Examples of osmosis

When it rains, the soil becomes wet. It has a high concentration of water particles. This concentration is often higher than that in the plant's root hair cells, so water moves by osmosis (or osmoses) into the plant. It moves from a high water concentration to a lower water concentration across the membrane of the root hair cells.

> **Exam tip**
>
> You should be able to describe how the factors in Table 1.4 affect diffusion.

> **Exam tip**
>
> You should be able to explain why exchange surfaces and a transport system are needed. In particular, you should be able to explain how the small intestine and lungs in mammals, gills in fish and the roots and leaves in plants are adapted.

> **Kinetic energy**: The store of movement energy.

> **Revision activity**
>
> Draw out this table with only the headings along the top and the first column on the left. Try to fill in the rest of the table from memory to help you to revise.

> **Osmosis**: The net diffusion of water from an area of high water concentration to an area of lower water concentration across a partially permeable membrane.
>
> **Partially permeable**: Allowing only substances of a certain size through.

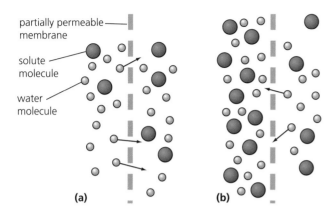

partially permeable membrane

solute molecule

water molecule

(a)　　(b)

Figure 1.14 Two partially permeable membranes. Both membranes have the same concentration of solution on the right-hand side. However, water moves in opposite directions through the membranes.

We can see osmosis when we put cells into different solutions. Your red blood cells and the plasma they are surrounded by are the same concentration. So, water can move from the cells into the plasma and in reverse, but no net movement occurs. Solutions with the same overall concentration are called isotonic.

If a red blood cell is put into a solution of salty water it will shrivel and shrink. There is a higher concentration of water in the cell than in the surrounding solution. So, the water will move by osmosis from the cell into the solution. If one solution has a higher concentration than another one, we call the first one **hypertonic**.

If a red blood cell is put into a solution of distilled water it will swell and could burst. There is a lower concentration of water in the cell than in the surrounding solution. So, the water will move by osmosis from the solution into the cell. If one solution has a lower concentration than another one, we call the first one **hypotonic**.

> **Hypertonic**: A solution with a higher concentration of solutes.
>
> **Hypotonic**: A solution with a lower concentration of solutes.

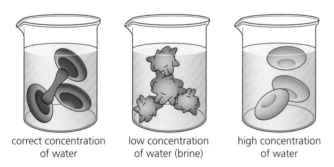

correct concentration of water　　low concentration of water (brine)　　high concentration of water

Figure 1.15 This is what happens to red blood cells in solutions with different concentrations of water. (Not to scale.)

Required practical 2

Investigating osmosis

Aim: To investigate the effect of a range of concentrations of salt or sugar solutions on the mass of plant tissue.

Equipment: Boiling tubes, five salt or sugar solutions, distilled water, potato discs.

Method:

1 Cut six small discs of potato and individually record the mass of each.
2 Place one disc in a boiling tube of each of the five different salt or sugar solutions.
3 Place one disc in a boiling tube of distilled water to act as a control.
4 Wait for 30 minutes.
5 Remove each disc and re-measure its mass.
6 Calculate the percentage change in mass for each disc.

Results:

Plot a graph of concentration of salt or sugar solution and change in mass. If the solution contains more sugar or salt than the potato disc, then the percentage change in mass will be lower than 100% because water moves from a high concentration of water in the disc to a lower concentration in the solution.

If the solution contains less sugar or salt than the potato disc, then the percentage change in mass will be greater than 100% because water moves from the high water concentration in the solution to the lower concentration in the disc.

If the disc stays the same mass, then the concentrations are the same and no net movement has occurred.

> **Exam tip**
>
> You should be able to calculate the rate at which water is absorbed into or lost from a plant, and calculate the overall percentage gain or loss of mass.

Active transport

REVISED

Active transport is the net movement of particles from an area of lower to higher concentration. So, active transport reverses the effects of diffusion.

This movement does not happen naturally and so requires the use of energy. It is therefore not a passive process like diffusion and osmosis, but an active process. Because active transport always happens from a lower to a higher concentration, we say that it occurs up a concentration gradient.

> **Active transport:** The net movement of particles from an area of lower concentration to an area of higher concentration using energy.

Examples of active transport

Water moves from a high water concentration in the soil to a lower water concentration in the plant by osmosis. This movement does not require energy. But what happens to mineral ions found in the soil that plants need to absorb? Minerals are present at higher concentrations in the plant than in the soil. So, they cannot move into the plant by diffusion. Plants need to absorb these mineral ions from the lower concentration in the soil to the higher concentration in the plant. This movement requires active transport, which uses energy.

This process also occurs in your digestive system. When you have just digested a meal, glucose is found at a high concentration inside your small intestine. It moves by diffusion into your blood. This does not require energy. But what happens when most of this glucose has been absorbed? The glucose is now at a higher concentration in your blood and so cannot be absorbed by diffusion. Your body absorbs the last of the glucose into your blood by active transport. This final movement uses energy.

> **Exam tip**
>
> You should be able to describe how substances are transported into and out of cells by diffusion, osmosis and active transport. In doing so, you should be able to explain the differences between the three processes.

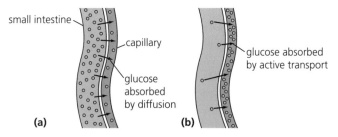

small intestine
capillary
glucose absorbed by active transport
glucose absorbed by diffusion

(a)   (b)

Figure 1.16 Look carefully at the glucose concentrations in the intestine and in the blood: a) diffusion, and b) active transport.

Revision activity

Draw a table with the headings diffusion, osmosis and active transport. In the table, describe each process (perhaps with a diagram) and give examples of where each occurs.

Now test yourself

TESTED ☐

20 Define the term diffusion.
21 In which two places does oxygen diffuse in the body?
22 Define the term osmosis.
23 How is osmosis different from diffusion?
24 What term do we give to a solution with a lower concentration of solutes?
25 By what process does water move into plant root hair cells?
26 Define the term active transport.
27 What does active transport require that diffusion does not?
28 What two processes help move glucose from the small intestine into the capillaries?
29 Why do plants absorb mineral ions from the soil by active transport?

Answers online

Summary

- Animal and plant cells are eukaryotic, so possess a nucleus. Bacterial cells are prokaryotic and so do not.
- Animal cells have a nucleus, cytoplasm, cell membrane, mitochondria and ribosomes. Plant cells have the same components as animal cells, as well as chloroplasts and a permanent vacuole filled with cell sap. Plant and algal cells have a cell wall made from cellulose.
- Bacterial cells are smaller than plant or animal cells. They have a cytoplasm, cell membrane and wall. Their DNA is found as a single loop. They may also have DNA plasmids.
- Cell specialisation allows cells to complete a specific function. Examples in animals include sperm, nerve and muscle cells. Examples in plants include xylem, phloem and root hair cells.
- As organisms develop their cells differentiate to become specialised. These cells develop different components to fulfil their function. Differentiation happens early in the life of most animals. Many plant cells can differentiate throughout their life.

- Microscope technology has developed over time. Electron microscopes have a much higher magnification and resolution than light microscopes. Magnification is calculated as the size of the image divided by the size of the real object.
- The nucleus of eukaryotic cells contains chromosomes made from DNA. Chromosomes contain large numbers of genes. In diploid cells, chromosomes exist as pairs.
- The life of a cell is described in the cell cycle. Mitosis is a part of this cycle. During mitosis, chromosomes double before the cell divides into two identical daughter cells. Mitosis is important in the growth and development of organisms.
- Stem cells are undifferentiated. In humans, embryonic stem cells can develop into many different cells types, whilst adult stem cells can only develop into a much smaller number of cell types. Treatment with stem cells, including therapeutic cloning, could help diabetic or paralysed patients. Stem cell research is an ethical issue.

→

- Plant stem cells are found in regions called meristems and can develop into any cell type. They can be used in tissue culture to produce clones of rare plants or of important crops.
- Diffusion is the net movement of particles from an area of high to lower concentration. Oxygen and carbon dioxide move in and out of cells by diffusion. The rate of diffusion is affected by the concentration gradient, temperature and surface area.
- Larger organisms require exchange surfaces such as lungs or gills to maximise diffusion.
- Osmosis is the net movement of water from a high to a lower concentration across a membrane. Water moves into roots and through plants by osmosis.
- Active transport is the net movement of particles from an area of low to higher concentration and requires energy. Plants absorb mineral ions by active transport. You absorb the last glucose molecules from your small intestine into your blood by active transport.

Exam practice

1 What is present in animal and plant cells but not bacterial ones? [1]
   A Cytoplasm
   B Cell membrane
   C Nucleus
   D Cellulose cell wall
2 Which type of cell is prokaryotic? [1]
   A Plant
   B Animal
   C Bacterial
   D Fungal
3 Define mitosis. [1]
4 How do human embryonic and adult stem cells differ? [2]
5 Describe the differences between light and electron microscopes. [2]
6 What is the total magnification of a light microscope with an eyepiece lens of ×10 and an objective lens of ×25? [1]
7 Describe how you would make a light microscope slide of a human cheek cell. [4]
8 Compare and contrast the processes of diffusion and active transport. [6]
9 Compare and contrast plant and animal cells. [6]

Answers and quick quiz 1 online

ONLINE

2 Organisation

Animal tissues, organs and systems

Levels of organisation in living organisms

Larger multicellular animals have several levels of organisation. From smallest to largest, they are:
1 Cells are the basic building blocks of all life.
2 Tissues are groups of cells with a similar structure and function.
3 Organs are groups of tissues that perform a specific function.
4 Organ systems are groups of organs with similar functions.
5 Organisms are made from organ systems.

Table 2.1 **Examples of levels of organisation.**

Organisational level	Examples
Cell	Nerve cell, muscle cell
Tissue	Nervous tissue, skin
Organ	Brain, heart
Organ system	Nervous system, digestive system
Organism	Human, frog

Revision activity

Draw out this table with only the headings along the top and the first column on the left. Try to fill in the rest of the table from memory to help you to revise.

The human digestive system

Your digestive system is about nine metres long and runs from your mouth to your anus.

It breaks down the large, **insoluble** bits of food that you eat. These are broken down into smaller, **soluble** pieces that can be absorbed into your blood. Once this happens they are transported around your body to the cells that need them.

Insoluble: Cannot dissolve.

Soluble: Can dissolve.

Functions of the parts of the digestive system

The locations of the parts of your digestive system are shown in Figure 2.1. The functions of these components are found in Table 2.2.

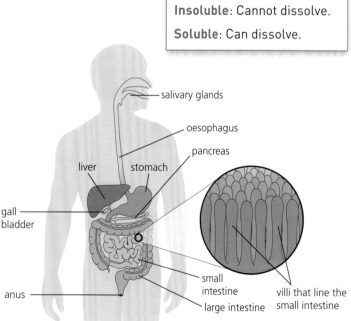

Figure 2.1 **The digestive system showing the location of the villi.**

Table 2.2 The parts of the digestive system and their functions.

Component	Function
Salivary glands	Salivary glands in your cheeks produce saliva. This saliva lubricates food as it passes along your oesophagus. Saliva also contains a carbohydrase **enzyme** called amylase that begins the break-down of carbohydrates into sugars.
Oesophagus	This short tube connects your mouth and stomach.
Stomach	Your stomach is a small bag about the size of your fist. It has ridges that allow it to increase in size when you eat food. Food is mixed with stomach acid to kill any **pathogens**. Stomach acid does not break down food. Protease enzymes are mixed with food to begin the break-down of proteins.
Liver	Food does not pass though the liver. The liver produces **bile**, which breaks down fats into smaller sections. This process is called emulsification. It increases the surface area of the food to allow lipase enzymes to work more effectively.
Gall bladder	Food does not pass through the gall bladder. Bile is stored in the gall bladder before being released into the small intestine.
Pancreas	Food does not pass through the pancreas. It produces carbohydrase, protease and lipase enzymes and releases these into the small intestine.
Small intestine	Digested food is absorbed into the blood in the small intestine. It is about six metres long. The surface of the small intestine is not smooth. It possesses millions of tiny finger-like projections called villi. These villi increase the surface area of the small intestine to allow more nutrients to be absorbed into the blood. Food is pushed through your small intestine by a process called **peristalsis**. This process is the rhythmical contraction and relaxation of muscles in the lining of the small intestine. This movement forces lumps of food along it.
Large intestine	All that is left of your food when it leaves the small intestine is water and fibre that you cannot digest. The large intestine absorbs water from this food, leaving fibre which forms your solid waste (faeces).
Anus	This opening controls when you release faeces when you go to the toilet.

Enzyme: A biological molecule that speeds up a chemical reaction.

Pathogen: A disease-causing micro-organism (bacterium, fungus or virus).

Bile: A green-coloured liquid produced by your liver, stored by your gall bladder and released into your small intestine to break down fats.

Peristalsis: The rhythmical contraction of muscle behind food in your digestive system to push it along.

Revision activity

Draw out this table with only the headings along the top and the first column on the left. Try to fill in the rest of the table from memory to help you to revise.

Revision activity

Pushing a tennis ball through a pair of tights is a good model for peristalsis pushing a bolus of food along your digestive system.

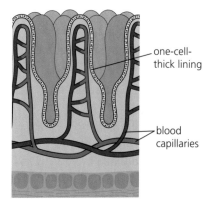

one-cell-thick lining

blood capillaries

Figure 2.2 Villi are small, hair-like structures in your small intestine. Villi increase the surface area over which molecules of digested food are absorbed.

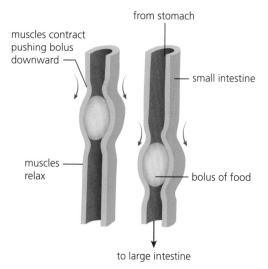

from stomach

muscles contract pushing bolus downward

small intestine

muscles relax

bolus of food

to large intestine

Figure 2.3 The rhythmical contraction and relaxation of the muscles that line much of the digestive system is called peristalsis.

Required practical 3

Food tests

Aim: To use qualitative reagents to test for a range of carbohydrates, **lipids** and proteins.

Equipment: Spotting tile, boiling tube, iodine solution, distilled water, Benedict's solution, water bath, Biuret solution, bung.

Method and results – starch test:
1 Place a small amount of food onto a spotting tile.
2 Add two drops of iodine solution.
3 If the food turns blue or black, starch is present.
   If it remains brown (the colour of iodine solution) then no starch is present.

Method and results – glucose test:
1 Place a small amount of food in a boiling tube.
2 Add 10 cm³ of distilled water.
3 Add 10 drops of Biuret solution to the boiling tube.
4 Heat in a water bath at 80 °C for 10 minutes.
5 If the solution turns orange or green, glucose is present.
   If it remains blue (the colour of Benedict's solution) then no glucose is present.

Method and results – protein test:
1 Place a small amount of food in a boiling tube.
2 Add 10 cm³ of distilled water.
3 Add 10 drops of Biuret solution to the boiling tube.
4 If the solution turns a light lilac colour, then protein is present.
   If it remains blue (the colour of Biuret solution) then no protein is present.

Method and results – oils test:
1 Place a small amount of food in a boiling tube.
2 Add 10 cm³ of distilled water.
3 Place bung in boiling tube and shake vigorously.
4 If an oil is present an emulsion will form and the water will turn cloudy.

Lipids: Fats or oils, which are insoluble in water.

Enzymes

Enzymes are biological catalysts. They speed up reactions and are not used up in them. This section focuses on the enzymes present in your digestive system. These enzymes break down large molecules of food into smaller ones. They are called break-down enzymes. There are other enzymes, however, that do the reverse. They join smaller molecules together to make larger ones. The enzyme involved in protein synthesis does this, for example. These are called synthesis enzymes.

> **Exam tip**
>
> You should be able to relate your knowledge of enzymes to metabolism.

Human digestive enzymes

There are three types of digestive enzyme. The molecules of food that they break down are called **substrates**. The three types of enzymes, their substrates and products and where they are found are shown in Table 2.3.

> **Substrate**: The molecule on which an enzyme acts.

Table 2.3 The enzymes, substrates and products of the digestive system.

Enzyme	Substrate	Product	Location
Carbohydrase	Carbohydrates	Sugars	Mouth, pancreas and small intestine
Protease	Proteins	Amino acids	Stomach, pancreas and small intestine
Lipase	Fats and oils (lipids)	Fatty acids and glycerol	Pancreas and small intestine

> **Revision activity**
>
> Draw out this table with only the headings along the top and the first column on the left. Try to fill in the rest of the table from memory to help you to revise.

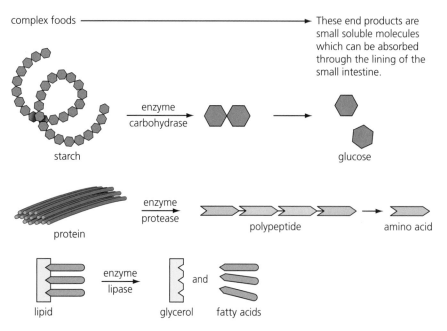

complex foods ⟶ These end products are small soluble molecules which can be absorbed through the lining of the small intestine.

starch → enzyme carbohdrase → → glucose

protein → enzyme protease → polypeptide → amino acid

lipid → enzyme lipase → glycerol and fatty acids

Figure 2.4 The break-down of complex food molecules into small, soluble molecules that can be used.

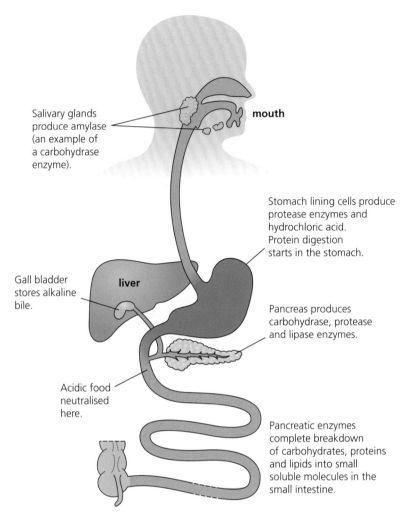

Salivary glands produce amylase (an example of a carbohydrase enzyme).

mouth

Stomach lining cells produce protease enzymes and hydrochloric acid. Protein digestion starts in the stomach.

Gall bladder stores alkaline bile.

liver

Pancreas produces carbohydrase, protease and lipase enzymes.

Acidic food neutralised here.

Pancreatic enzymes complete breakdown of carbohydrates, proteins and lipids into small soluble molecules in the small intestine.

Figure 2.5 Digestive enzymes control reactions that take place in the digestive system. No enzymes are made or used in the oesophagus, liver (bile is not an enzyme), gall bladder, large intestine or anus.

Bile

Bile is not an enzyme. It does not break down lipids into fatty acids and glycerol as lipase enzymes do. Bile is an emulsifier. It breaks down large globules of fat into smaller ones. This process increases the surface area that lipase enzymes can then work on. This process speeds up their digestion.

Bile is also an alkaline substance. It neutralises any excess stomach acid at the beginning of the small intestine. This process provides the enzymes in the small intestine with their optimum pH.

The lock and key theory

Enzymes are specific for their substrates like keys are specific for their locks. So, protease enzymes will not break down lipids, just as the key to your house will not open your parent's car. For an enzyme to break down a substrate, the substrate must fit into the enzyme, just like a key fits into a lock. So, the shape of the enzyme and substrate must match, just like keys and locks. This model is called the **lock and key theory**.

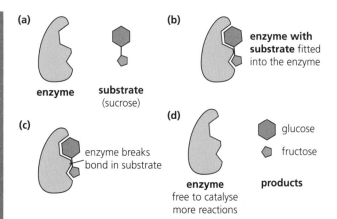

Figure 2.6 How a digestive enzyme breaks down a substrate. Here the substrate is sucrose and the products are glucose and fructose.

At optimum pH and temperature, the shapes of the enzyme and substrate fit together perfectly. When we move away from the optimum pH or temperature, the shape of the **active site** changes. This change makes it harder for the enzyme and substrate to fit together and so slows the rate at which the enzyme works. This in turn slows the reaction. If extremes of pH or temperature are reached, the shape of the active site is permanently changed. The enzyme's active site becomes **denatured** and will no longer function.

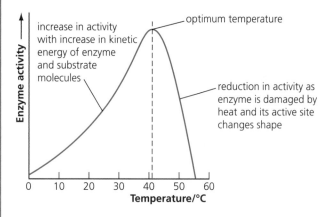

Figure 2.7 This graph shows the effect of temperature on the activity of an enzyme.

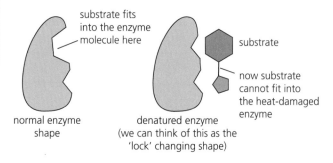

Figure 2.8 Extremes of temperature and pH denature enzymes by altering the shape of their active site, so the substrate can no longer fit.

> **Exam tip**
>
> You should be able to use the lock and key theory to explain how enzymes work.

> **Active site**: The region of an enzyme that binds to its substrate.
>
> **Denatured**: A permanent change to an enzyme as a result of extremes of pH and temperature that stop it working.

> **Exam tip**
>
> You should be able to relate the activity of enzymes to their temperature and pH.

> **Exam tip**
>
> You should be able to calculate the rate at which chemical reactions occur.

> **Exam tip**
>
> As well as the lock and key theory, enzyme activity at higher temperatures can be explained by particle theory. At higher temperatures, molecules have more kinetic energy so move faster. This means that they are more likely to collide with substrates.

Required practical 4

pH and rates of reaction

Aim: To investigate the effect of pH on the rate of reaction of amylase enzyme.

Equipment: Boiling tubes, iodine solution, amylase solution, pH buffer solutions, water bath, pipettes, spotting tile.

Method:

1 Place one drop of iodine into each well of a spotting tile.
2 Put 10 cm³ of starch solution in a boiling tube.
3 Place 2 cm³ of amylase solution in another boiling tube.
4 Add 5 cm³ of pH buffer solution to the second boiling tube.
5 Place both boiling tubes into a water bath at 37 °C for 2 minutes.
6 After 2 minutes, add the contents of both boiling tubes together.
7 Every 30 seconds, use a pipette to put one drop of solution into a new well of the spotting tile.
8 Repeat every 30 seconds until the solution turns blue or black.
9 Repeat with different pH buffer solutions.

Results: The longer the time that the iodine test gave positive results, by turning blue or black, the less suitable the pH is to the enzyme. So, the optimum pH is the one at which the solution remained brown for the longest.

> **Vein**: A large blood vessel that returns blood to the heart.
>
> **Artery (plural arteries)**: A large blood vessel that takes blood from the heart.

The heart and blood vessels

REVISED

Your circulatory system is made from your heart and all your **veins**, **arteries** and **capillaries**. Its function is to provide all the millions of billions of cells in your body with the substances they need. All cells need glucose and oxygen for respiration. All cells need carbon dioxide and water (the products of respiration) to be removed. Other cells require different substances such as **hormones** at specific times, such as during puberty. Your circulatory system ensures that all your cells have all the substances they need.

> **Capillaries**: Tiny blood vessels found between arteries and veins that carry blood into tissues and organs.
>
> **Hormone**: A chemical (produced in a gland in mammals) that moves around an organism to change the function of target cells, tissues or organs.

The heart and double circulation

Your heart is an organ made from muscle and nervous tissue. It pumps roughly every second and so your heart beats an amazing 100 000 times a day or over two billion times in your lifetime. Amazingly, many other mammals have heartbeats that are roughly similar and tend to beat around one billion times in their lifetime.

Your heartbeat is controlled by your 'natural pacemaker'. This is a small group of nervous tissue cells in the top right chamber

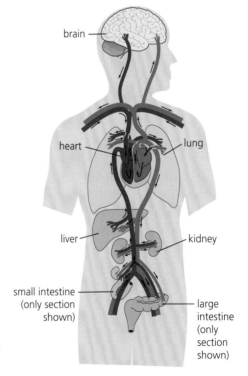

Figure 2.9 The circulatory system, showing the main organs with which the blood exchanges substances. A bright red colour indicates oxygenated blood and a dark red colour indicates deoxygenated blood.

of your heart. They generate an electrical signal which spreads out along your heart's nerve fibres and causes the heart muscle to contract. This action pumps blood from your heart. Your pacemaker controls the rate at which your heart beats.

Blood takes about 1 minute to complete a full circuit of your body. During this time, it will pass through your heart twice. Blood is pumped in a circle from your heart to your lungs, back to your heart and then to the rest of your body, before returning to your heart to start again. This is called double circulation.

Your heart has four chambers: two on each side. The top two chambers are called atria, or your left **atrium** and your right atrium. The bottom chambers are called **ventricles**. Atria and ventricles are separate by valves. Blood collects in your atria when the valves are closed. When your heart beats, the blood in your atria is forced into the ventricles. More valves at the ends of the ventricles stop blood being pumped straight into your blood vessels. When your heart beats again, the blood in your ventricles is forced into your blood vessels to begin another journey around your body.

> **Atrium (plural atria)**: An upper chamber of the heart surrounded by a smaller wall of muscle.
>
> **Ventricle**: A lower chamber of the heart surrounded by a larger wall of muscle.

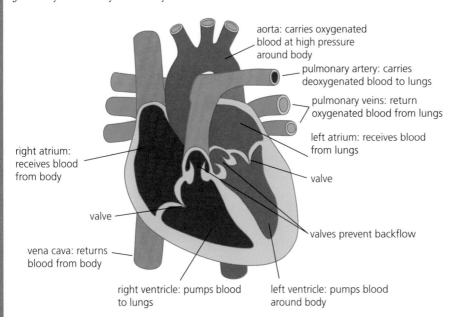

aorta: carries oxygenated blood at high pressure around body

pulmonary artery: carries deoxygenated blood to lungs

pulmonary veins: return oxygenated blood from lungs

left atrium: receives blood from lungs

valve

valves prevent backflow

right atrium: receives blood from body

valve

vena cava: returns blood from body

right ventricle: pumps blood to lungs

left ventricle: pumps blood around body

Figure 2.10 A cross-section of the heart, showing its structure.

> **Typical mistake**
>
> Convention dictates that labelled heart diagrams in all books (including this one) have the right-hand side on the left and the other way around.

Blood is pumped from your left atrium into your left ventricle. It is then pumped into an artery called the aorta. The aorta divides into smaller arteries and eventually into tiny capillaries that permeate your tissues and organs to provide oxygen, glucose and other required substances. Your capillaries collect into small veins that join to become larger and eventually form one vein called the vena cava. This vein returns blood to your right atrium. From here it is pumped into your right ventricle and then into an artery called the pulmonary artery. This artery takes blood to your lungs to replenish the oxygen and remove carbon dioxide. Your blood completes its journey by returning to the left atrium in your pulmonary vein.

The muscular lining of your heart is thicker on the left-hand side. This difference is because the left ventricle pumps blood to all tissues of your body including those in your extremities. The right ventricle only pumps blood to the lungs so does not need such a thick muscular lining.

> **Exam tip**
>
> You should be able to describe the structure and function of the human heart and lungs including their adaptations.

The blood vessels

You have between 50 000 and 100 000 miles of blood vessels in your body, which is enough to circle the Earth. There are three types: arteries, veins and capillaries.

Arteries take blood away from the heart. This blood is under high pressure, so the lining of arteries needs to be thick and muscular. It is also elastic so it can stretch when blood is pumped from your heart. You can feel this surge of blood in places such as your wrist where arteries are near the surface. This surge is called your pulse.

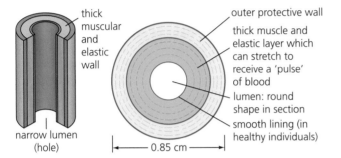

Figure 2.11 Note the thickness of the artery wall.

Capillaries carry blood into your tissues and organs to provide all your cells with the substances they need. Capillaries are very small and spread out into your tissues and organs like the roots of a tree. They are very thin. Blood plasma passes from capillaries into the tissues. Here plasma is called tissue fluid and it provides cells with glucose and oxygen for respiration and other substances. Glucose and oxygen move into the cells by diffusion. The products of respiration are carbon dioxide and water. Carbon dioxide and water diffuse into the tissue fluid to be carried away in the capillaries.

Veins carry blood back to the heart under much lower pressure. Blood loses pressure during its journey through the capillaries. So, the linings of your veins do not need to be as muscular as the linings of your arteries. Veins also have one-way valves (not found in arteries) to keep blood flowing back to your heart.

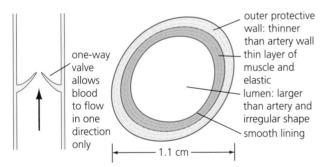

Figure 2.12 Note the irregular shape and the thinner muscle and elastic layer of the vein.

Typical mistake

The same volume of blood that leaves your heart must return to it. So even though images of veins and arteries show them to have differently shaped structures, the total area of the arteries leaving and the veins returning blood must be the same.

Exam tip

You should be able to explain how the structure of blood vessels relates to their function.

Exam tip

You should be able to calculate the rate of flow of blood.

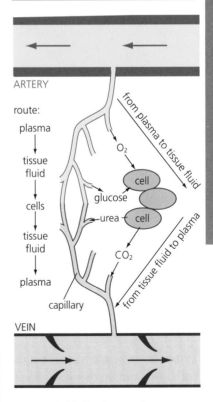

Figure 2.13 Exchange between the blood and tissue cells in a capillary network.

Components of blood

You have about 5 litres of blood in your body. Blood looks red because it carries millions of red blood cells around your body. Blood also carries other cells and substances including those absorbed into your blood by your digestive system. These substances include glucose, fatty acids, glycerol and amino acids.

Red blood cells

There are millions of red blood cells in every drop of your blood. These cells carry oxygen from your lungs to your tissues and organs where it is needed for respiration. Red blood cells are a characteristic **biconcave** shape. (They have dips in their middle on both sides.) This shape increases the surface area of the cell, and so increases its ability to absorb oxygen.

Red blood cells contain a compound called **haemoglobin**. In your lungs, this binds with oxygen to form **oxyhaemoglobin**. This process turns the colour of your blood from dark red to bright red. When these red blood cells reach your tissues, they release the oxygen and so oxyhaemoglobin turns back to haemoglobin. Red blood cells don't have a nucleus. This allows them to maximise the amount of haemoglobin they can carry.

Figure 2.14 The biconcave shape of red blood cells maximises their surface area to volume ratio and so increases their ability to absorb oxygen.

Biconcave: Describes a shape with a dip that curves inwards on both sides.

Haemoglobin: The molecule in red blood cells that can temporarily bind with oxygen to carry it around your body.

Oxyhaemoglobin: A substance formed when haemoglobin in your red blood cells temporarily binds with oxygen.

Phagocyte: A type of white blood cell that engulfs and destroys pathogens.

Lymphocyte: A type of white blood cell that produces antibodies to help clump pathogens together to make them easier to destroy.

Antigen: A protein on the surface of a pathogen that your antibodies can recognise as foreign.

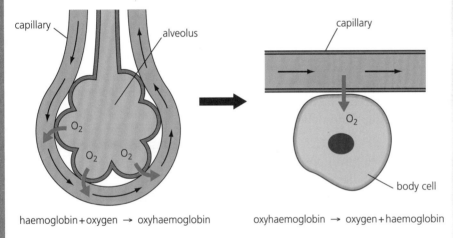

haemoglobin + oxygen → oxyhaemoglobin   oxyhaemoglobin → oxygen + haemoglobin

Figure 2.15 Haemoglobin transports oxygen from the lungs to other organs as oxyhaemoglobin in red blood cells.

White blood cells

White blood cells are part of your immune system to attack pathogens before they can make you ill. You have far fewer white blood cells than red ones. A single drop of your blood will only contain tens of thousands of white blood cells.

There are several types of white blood cells, but two main groups:

Phagocytes engulf pathogens by surrounding them and absorbing them. Break-down enzymes, present inside phagocytes, destroy pathogens. Phagocytes act non-specifically.

Lymphocytes make antibodies that bind to protein **antigens** found on pathogens and, in doing so, clump them together. This process makes it easier for pathogens to be destroyed by phagocytes. Each lymphocyte makes a specific type of antibody that binds to a specific protein antigen on the pathogen. Therefore different lymphocytes are specific to different pathogens.

Platelets

These are not cells, but fragments of them. There are over a 100 000 of them in a drop of blood. They join together after you have cut yourself and form a scab. When you cut yourself, **platelets** release clotting factors. These factors convert a chemical called fibrinogen in your blood into fibrin. This forms a mesh and sticks platelets to it to form a scab.

Plasma

Blood plasma is the straw-coloured fluid that carries red and white blood cells, platelets and all other substances around your body. Just over half of your body's total blood volume is plasma and nearly all your plasma is made of water.

Coronary heart disease: a non-communicable disease

REVISED

Communicable diseases like the common cold can be transmitted between people. Coronary heart disease is a non-communicable disease. This means it develops, rather than being caught.

Your heart is like all other organs in your body. Its cells need glucose and oxygen for respiration. These are delivered to your heart in its **coronary arteries**.

Unhealthy lifestyles include not doing enough exercise, smoking, drinking in excessive amounts and a poor diet. This can result in the build-up of fat inside coronary arteries and also a reduction in the flexibility of their linings. This process is called **atherosclerosis**. Fat build-up and atherosclerosis slow or stop blood from reaching the heart. This can result in heart attacks.

Coronary heart disease is now one of the major causes of death in the world. A common previous treatment for this was a **heart bypass** operation. This treatment involved moving a short section of artery from another part of a person's body and using it to short circuit the blockage. This is a major form of surgery and has inevitable risks. More recent treatments are less damaging. **Stents** are small medical meshes that can be expanded to hold open blocked arteries. Statins are drugs prescribed by doctors to reduce **cholesterol**, which in turn reduces coronary heart disease.

> **Platelets**: Small structures (not cells) in your blood that fuse together to form a scab.

> **Exam tip**
>
> You should be able to describe the functions of all blood components. You should also be able to recognise the different types of blood cell in a diagram and explain how their structure is related to their function.

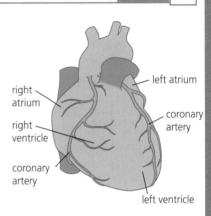

Figure 2.16 The coronary arteries supply glucose and oxygen to the heart muscle tissue.

> **Communicable**: A disease that can be transmitted from one organism to another.
>
> **Coronary arteries**: Arteries that supply the heart muscle with oxygenated blood.
>
> **Atherosclerosis**: A medical condition resulting from an unhealthy lifestyle that reduces the flexibility of arteries.
>
> **Heart bypass**: A medical procedure in which a section of less important artery is moved to allow blood to flow around a blockage in a more important one.
>
> **Stent**: A small medical device made from mesh that keeps arteries open.
>
> **Cholesterol**: An important biological molecule for cell membranes, which leads to atherosclerosis if found in high levels in the blood.

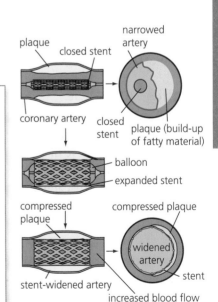

Figure 2.17 This stent allows blood to flow freely again.

Faulty valves

Your heart has valves between the atria and ventricles and also between the ventricles and arteries. Some heart valves do not function properly. If they do not function properly, less blood will be pumped around a person's body. Valves can be replaced in a major heart operation. They can be from organ donors or artificial, mechanical values.

Pacemakers

Some people have difficulties controlling their heartbeat. The beat needs to be coordinated properly to ensure the maximum volume of blood is pumped at each heartbeat. An artificial pacemaker can be fitted to people in this situation. The pacemaker takes over the generation of the electrical signal needed to coordinate every heartbeat.

Transplants

If all other treatments fail, some patients need a heart transplant. These treatments are serious medical procedures. All transplants, including this type, need a match between the donor and that of the patient. Some patients can sadly be on waiting lists for a heart transplant for many years.

> **Exam tip**
>
> You should be able to evaluate the advantages and disadvantages of the various cardiovascular treatments listed above.

Health issues

REVISED

Your health is now defined as your mental and physical wellbeing. It can be improved by:
- seeking medical help when you are physically or mentally unwell
- eating a balanced diet
- exercising regularly
- reducing stress levels.

Well-balanced diet

A well-balanced diet has the correct amounts of all food groups. This can be shown in a food pyramid. This model shows that a large part of your diet should be made from carbohydrates like bread, potatoes and rice. Another large part of your diet should be made from fruits and vegetables. These contain natural healthy sugars, vitamins and fibre. Fats are found in dairy products including milk and cheese. These are also high in protein. They do however, have high cholesterol levels and so should be eaten in smaller quantities.

Regular exercise

Exercise strengthens your muscles including your heart. It also improves the efficiency of your circulatory system. Many people also believe that exercise helps your mental, as well as physical, health. The National Health Service (NHS) recommends that all young people do an hour of exercise every day.

Figure 2.18 A balanced food diet represented as a food pyramid.

Physical and mental ill health

Doctors now believe that there are links between diseases and mental and physical health. Patients with one medical condition can be more likely to develop another:
- If a person's immune system is not functioning properly, they are more likely to catch communicable diseases.
- A small number of viral infections leads to cancer.
- Serious physical health problems can lead to mental ill health.

Stress is the feeling of being under too much mental or emotional pressure. Anxiety is the feeling of unease, which is often worry or fear. Depression affects different people in many ways including physical ones like losing appetite. It is crucial that anyone who feels stressed, anxious or depressed speaks to their doctor as soon as they can.

Cancer

Sometimes the process of cell division goes wrong and forms cancerous cells. These multiply in an uncontrolled way and can form a lump or tumour. **Malignant** tumours are cancerous. They can spread to other parts of the body and form secondary tumours in a process called **metastasis**. It is essential that malignant tumours are removed in an operation or that their cells are killed before they spread. **Benign** tumours are less serious. They often are found within a membrane and so are less likely to spread.

Sadly, it is likely that one in every two people will suffer from cancer at some point in their life. Symptoms include a lump, weight loss and unexplained bleeding. There are over 200 types of cancer, many of which have different symptoms. It is crucial that anyone with concerns speaks to their doctor as soon as they can.

Screening

Screening is the medical process of looking for cancers. Doctors feel for lumps, take blood and urine tests and X-rays during screening. Women often have two additional screening tests. **Mammograms** are X-rays that are automatically offered by the NHS to women between 50 and 70 years of age to screen for breast cancer. **Smear tests** look for cervical cancer. Cells are collected from the **cervix** and are viewed under a microscope to check for cancerous cells. The NHS automatically offers smear tests to women between 25 and 49 years of age every 2 years, and less frequently beyond the age of 50.

Causes of cancer

More than 20% of cancers are caused by smoking. Excessive drinking also causes many cancers. Hepatitis B and C virus infections and human papilloma virus (HPV) can also cause cancer. **Ionising radiation** from the Sun's ultraviolet (UV) rays and chemical pollutants from factories can cause cancer. The older a person gets the more likely they are to suffer from cancer.

Treating and preventing cancer

Chemotherapy and radiotherapy are the two most common forms of cancer treatment. They are often used together. Chemotherapy uses chemicals to kill cancerous cells. Radiotherapy uses X-rays to do the same. With both treatments, it is often difficult for doctors to only kill cancerous cells. Side effects of both treatments include hair loss, being tired and being sick.

The development of cancers is prevented by leading a healthy lifestyle. This includes not smoking, drinking in low amounts or not at all, exercise and a balanced diet.

The effect of lifestyle on some non-communicable diseases

Risk factors are aspects of your lifestyle or substances taken into your body that increase the risk of a disease occurring. Some risk factors are known to cause disease. We call this

---

Exam tip

You should be able to describe cancer as the result of changes in cells that lead to uncontrolled growth and division.

Malignant: A cancerous tumour that can spread to other parts of the body.

Metastasis: The development of tumours following the spread of malignant cancer.

Benign: A non-cancerous tumour that does not spread.

Mammogram: A medical procedure using X-rays to check for breast cancer.

Smear test: A medical procedure to check for cancer of the cervix.

Cervix: The narrowing between the vagina and the uterus in the female reproductive system.

Ionising radiation: UV rays, X-rays and gamma rays that can cause **mutations** to DNA.

Mutation: A permanent change to DNA, which may be advantageous, disadvantageous or have no effect.

causation. Others are only associated with diseases. Here no cause had been proved. This is called **correlation**. Table 2.4 shows risk factors, diseases and their effects.

Table 2.4 Risk factors, diseases and their effects.

Risk factor	Disease	Effects
Obesity and lack of exercise	Type 2 diabetes	Body cells do not respond to the hormone insulin, which helps control the glucose level in the blood.
Alcohol	Liver function	Long-term alcohol use causes liver cirrhosis. The cells in the liver stop working and are replaced by scar tissue. This stops the liver from removing toxins, storing glucose as glycogen and making bile.
Alcohol	Brain function	Excessive use of alcohol can also alter the chemicals in the brain (neurotransmitters), which pass messages between nerve cells. This can cause anxiety and depression, and reduced brain function.
Smoking	Lung disease and cancer	Smoking can cause cancer in many parts of the body, including the lungs, mouth, nose, throat, liver and blood. It also increases the chances of having asthma, bronchitis and emphysema.
Smoking and alcohol	Underdevelopment of unborn babies	Alcohol and chemicals from cigarettes in the mother's blood pass through the placenta to her baby. Without a fully developed liver the baby cannot detoxify these as well as the mother can. This can lead to miscarriage, premature birth, low birth weight and reduced brain function.
Carcinogens and ionising radiation	Cancer	Chemicals and radiation that cause cancer are called **carcinogens**. Tar in cigarettes, asbestos, ultraviolet from sunlight and X-rays are examples.

Exam tips

- You should be able to describe the relationship between health and disease and the interactions between diseases shown in Table 2.4.
- You should be able to describe data about diseases in frequency tables and diagrams, bar charts and histograms, and use a scatter diagram to identify correlations.
- You should be able to discuss the human and financial cost of non-communicable diseases.
- You should be able to explain the effect of lifestyle factors on the levels on non-communicable disease.

Risk factor: Any aspect of your lifestyle or substance in your body that increases the risk of a disease developing.

Causation: The act of causing an outcome.

Correlation: When an action and outcome are linked, but the action does not cause the outcome.

Now test yourself

TESTED

1 Describe the process of peristalsis.
2 What are the three enzymes of the human digestive system?
3 Describe the food test for starch.
4 Describe what occurs when an enzyme is denatured.
5 Why is the human circulatory system often described as double circulation?
6 How are veins adapted for their function?
7 What are the four components of blood?
8 What is the function of lymphocytes?
9 Why are stents used more commonly for coronary heart disease than bypass operations?
10 How can you improve your health?

Answers online

Revision activity

Draw out Table 2.4 with only the headings along the top and the first column on the left. Try to fill in the rest of the table from memory to help you to revise.

Plant tissues, organs and systems

Plant tissues

Just as with you and other animals, multicellular plants have several levels of organisation. Cells are the basic building blocks of all life. Tissues are groups of cells with a similar structure and function.

Epidermal tissue

The outside layer of a plant is called its **epidermis**. This layer of cells has many functions including protection against the loss of water, the exchange of gases between the leaves and the air and the uptake of water in the roots. The epidermis is transparent to allow light to pass through it for **photosynthesis**.

Palisade mesophyll

Below the epidermis in leaves is the **palisade mesophyll** layer. This layer contains cells with very high numbers of chloroplasts to maximise the amount of glucose produced in photosynthesis.

Spongy mesophyll

Below the palisade mesophyll layer in leaves is the **spongy mesophyll** layer. These cells have fewer chloroplasts because they are further from sunlight. The cells of this layer are less regularly shaped and have gaps in between them. These gaps allow gases to diffuse from within the leaf into the air and the reverse.

Small pores called **stomata**, which are mainly found on the bottom of leaves, open and close to allow gases, including water vapour, to diffuse into and out from leaves at different rates. Stomata are surrounded by guard cells that swell to open the stomata and shrink to close it.

Xylem and phloem

Xylem tubes carry water and mineral ions from the roots to the leaves in a process called **transpiration**. Xylem tissue is made from hollow tubes strengthened by lignin to transport water to the leaves for photosynthesis.

The sugars made in photosynthesis dissolve in water and are transported around the plant in phloem tubes for respiration or storage in a process called translocation. Phloem tissues are made from tubes of elongated cells. Cell sap can move from one cell to the next through pores.

> **Palisade mesophyll**: Cells found towards the top of leaves with lots of chloroplasts for photosynthesis.
>
> **Spongy mesophyll**: Cells found towards the bottom of leaves with spaces in between them to allow gases to diffuse.
>
> **Stomata (singular stoma)**: Tiny holes in leaves bordered by guard cells that allow gases to diffuse in and out.
>
> **Xylem**: Dead plant cells joined together into long tubes through which water flows during transpiration.
>
> **Transpiration**: The gradual release of water vapour from leaves to continue the 'pull' of water up to them from the soil.

> **Epidermis**: The outermost layer of cells of an organism.
>
> **Photosynthesis**: A chemical reaction that occurs in the chloroplasts of plants and algae to store energy in glucose.

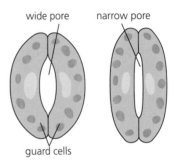

wide pore   narrow pore

guard cells

Figure 2.19 Pores called stomata become smaller if a plant needs to reduce water loss by transpiration.

upper side of leaf

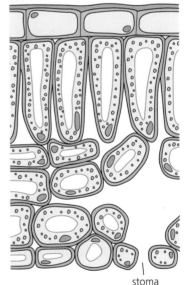

stoma

Figure 2.20 Part of a leaf in cross-section showing the epidermis (top and bottom layers), palisade mesophyll cells (long, thin upright cells below the upper epidermis) and the spongy mesophyll cells (more circular cells found towards the bottom of the leaf).

Xylem and phloem tubes are often found together in structures called vascular bundles.

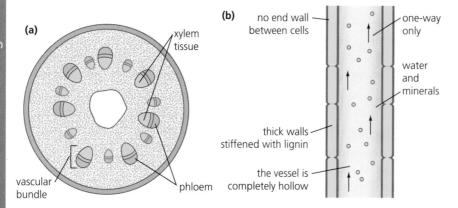

Figure 2.21 (a) Transverse section of a stem, (b) longitudinal section of xylem vessel from a vascular bundle (the arrows show the direction of water flow).

Meristem

The tips of roots and shoots have a region of rapidly growing cells called the meristem. Here mitosis happens.

Plant organs

REVISED

Just like in you and other animals, plant organs are groups of tissues that perform a specific function.

Root

Roots grow downwards towards water. They are usually white because they possess no chloroplasts. Cells that spend their lives underground do not need to photosynthesise. Roots anchor plants in the soil, between rocks or on trees. Some roots store excess glucose made in photosynthesis as starch. Root hair cells have an extension protruding into the soil. This massively increases their surface area, meaning they can absorb more water by osmosis and minerals by active transport.

Shoot

A shoot is the newly formed stem, its leaves and any buds. Just like the tips of roots, shoots also possess a meristem. Here cells divide by mitosis so the shoot continues to grow towards the light. (Root meristems allow roots to continue to grow downwards towards water.)

Leaf

Leaves are plant organs. They are the major site of photosynthesis. Evergreen plants retain their leaves throughout the year. **Deciduous** plants drop their leaves in autumn when light levels are too low to keep them. They regrow in the spring when light levels increase again.

Deciduous: Describes broadleaved trees that drop their leaves in winter.

Just like in you and other animals, plant organ systems are groups of organs with similar functions. Plants are made from their organ systems.

Transportation organ system

The transportation organ plant system is made from the roots, shoots and leaves of plants. (These are all organs.) This system acts like your circulatory system and moves all the substances the plant needs around their structures.

Translocation

Phloem tubes carry sugars and minerals dissolved in water throughout a plant. Sugars are made during photosynthesis and are broken down to release energy during respiration.

Transpiration

Plants do not have a heart so they cannot pump water to their leaves for photosynthesis. Instead we say it is 'pulled' upwards in a transpiration stream. Water is purposefully allowed to evaporate from the stomata. Guard cells control how quickly or slowly this happens. As the water evaporates, more is pulled up from the roots to replace it. This process therefore occurs continuously throughout the life of all plants.

Transporting water by transpiration allows plant cells to remain **turgid**. Transpiration is vital for photosynthesis and also enables essential minerals to be carried from the soil.

Transpiration rates increase when:
- There is more wind.
- The air is drier (less **humid**).
- The temperature is higher.
- The light intensity is higher (on sunny days).

> **Exam tip**
>
> You should be able to calculate the rate of transpiration. You should then be able to explain the effect of changing temperature, humidity, air movement and light intensity on the rate of transpiration.

> **Exam tip**
>
> You should be able to describe the processes of transpiration and translocation, including the structure of stomata.

> **Turgid**: Describes swollen cells.
>
> **Humid**: Describes an atmosphere with high levels of water vapour.

Now test yourself

TESTED

11 What are plant tissues?
12 What cell component is present in high numbers in the palisade mesophyll cells and why?
13 Why are there spaces between cells in the spongy mesophyll layer?
14 When a plant is respiring, what gases move out from the leaves?
15 Where are guard cells found?
16 Describe the process of transpiration.
17 Describe the process of translocation.
18 What term do we use to describe both xylem and phloem tubes close together in a plant?
19 What is different about xylem and phloem tubes?

Answers online

Summary

- Cells are the basic building blocks of all living organisms. Tissues are groups of cells with a similar structure and function. Organs are groups of tissues performing specific functions. Organs are arranged into organ systems which make up organisms.
- The digestive system is an organ system that breaks down large lumps of insoluble food, into smaller, soluble molecules, which are absorbed into the blood.
- Enzymes are biological catalysts that speed up reactions. An enzyme's substrate fits into its active site. The lock and key theory is a model to explain that enzymes are specific for their substrates.
- Enzymes have an optimum pH and temperature at which their activity is greatest. At extremes of pH or temperature, the shape of the active site changes and the enzyme becomes denatured. The substrate will no longer fit and so the enzyme will no longer work.
- Carbohydrase enzymes break down carbohydrates into sugars. Lipase enzymes break down fats and oils (lipids) into fatty acids and glycerol. Protease enzymes break down proteins into amino acids. The pancreas and small intestine make all three enzymes. The mouth makes carbohydrase enzymes. The stomach makes protease enzymes. Bile is not an enzyme, but it emulsifies fats. It is made in the liver.
- The human heart pumps blood around the body in a double circulation system. The heart is made from four chambers. The atria (upper chambers) pump blood to the ventricles, which then pump blood either to the lungs (right ventricle) or to the rest of the body (left ventricle).
- The lungs provide the blood with oxygen and remove carbon dioxide.
- Arteries, like the aorta and the pulmonary artery, take blood away from the heart. Arteries divide into tiny capillaries, which permeate tissues. Veins, like the vena cava and the pulmonary vein, return blood to the heart.

- Blood is a tissue. It contains red blood cells that carry oxygen, white blood cells that are part of the immune system and platelets that are cell fragments that form scabs. These are all carried in blood plasma.
- Coronary heart disease occurs when layers of fat block the inside of the arteries supplying the heart. The fat reduces the supply of oxygen and glucose needed for respiration. Bypass operations, the insertion of stents and statin drugs are all used to treat coronary heart disease. Heart and lung transplants can replace faulty organs.
- Health is the state of physical and mental wellbeing. Diseases are a major cause of ill health. Other factors like diet, stress and life situations can cause poor physical and mental health. Different types of diseases can interact.
- Risk factors are linked to an increased rate of a disease. Risk factors can be part of a person's lifestyle or substances in the body or environment. Causation occurs when risk factors have a proven link to disease. An example of this is smoking and lung disease and cancer.
- Cancer is the result of changes in cells that lead to rapid cell growth. Benign tumours are growths of abnormal cells that do not spread. Malignant tumours are cancers that do spread.
- Plant tissues include epidermal tissues, palisade mesophyll, spongy mesophyll, xylem and phloem and meristem tissue at the tips of roots and shoots. The leaf is a plant organ.
- Transpiration is the movement of water from the soil through root hair cells, up xylem tubes to the leaves. Plants release water vapour through stomata in their leaves to ensure transpiration occurs continuously. Transpiration rate is affected by temperature, humidity, air movement and light intensity.
- The roots, stem and leaves form the plant transportation organ system.
- Translocation is the movement of dissolved sugars made during photosynthesis from the leaves to the rest of the plant in phloem cells.

Exam practice

1 What type of enzyme is present in saliva? [1]
 A Carbohydrase
 B Lipase
 C Protease
 D Bile
2 What name is given to the process of moving sugary water from the leaves of a plant? [1]
 A Transpiration
 B Photosynthesis
 C Osmosis
 D Translocation
3 Define the term artery. [1]
4 Describe the pathway of blood around the body beginning with the left atrium. (Specific names of blood vessels are not required.) [4]
5 Describe the hierarchy of organisation in plants. [4]
6 Describe the shape of this graph and explain its significance. [4]

7 Describe how you investigate the effect of pH on enzyme activity of carbohydrase enzyme. [6]
8 Explain how plant roots are adapted for osmosis and active transport and give an example of a substance absorbed by each process. [6]
9 Explain how the lock and key theory models enzyme action including denaturing. [6]

Answers and quick quiz 2 online

ONLINE

3 Infection and response

Communicable (infectious) diseases

Pathogens are micro-organisms that pass disease from one organism to another. Diseases which can be passed by pathogens are called communicable (or infectious). There are four types of pathogen:
- viruses, such as measles
- bacteria, such as *Salmonella*
- fungi, such as rose black spot in plants
- protists, such as **malaria**.

The lifecycle of all pathogens is similar:
1 They infect a host.
2 They reproduce (or replicate if a virus).
3 They spread from their host.
4 They infect another host and repeat.

Malaria: A communicable disease, caused by a protist transmitted in mosquitos, which attacks red blood cells.

Spread of pathogens

REVISED

Pathogens spread disease in the following ways:
1 Airborne: when people sneeze, the virus that causes the common cold can be spread through the air in tiny droplets.
2 Direct contact (sexual or non-sexual): the *Chlamydia* bacterium is the cause of a sexually transmitted disease (STD) that passes from one person to another during sex.
3 Dirty water: cholera is caused by a bacterium that is spread in dirty water.
4 Contaminated food: food poisoning is often caused by the *Escherichia coli* bacterium in undercooked or reheated food.
5 **Vectors**: some farmers believe that the tuberculosis bacterium is passed from badgers to their cows. Any animal, or micro-organism that does this is called a vector.

The spread of disease can be reduced or prevented by:
- high levels of personal hygiene including hand washing
- covering your mouth and nose when you cough or sneeze
- cleaning and disinfecting surfaces and objects with antiseptics
- **vaccination** and taking medicines when prescribed
- avoiding close contact with people who are sick.

Exam tip

You should be able to explain how diseases caused by viruses, bacteria, protists and fungi are spread in animals and plants. You should also be able to explain how this spread can be reduced.

Vector: An animal that transmits a communicable disease without being infected itself.

Vaccine: A medicine containing an antigen from a pathogen that triggers a low level immune response so that subsequent infection is dealt with more effectively by the body's own immune system.

Viral diseases

REVISED

Viruses are not considered to be alive because they do not fulfil all the seven life processes. For example, they replicate not reproduce and they do not respire. They are therefore classified as strains and not **species**. They are some of the smallest pathogens, made from short lengths of DNA or RNA surrounded by a protein coat. They infect individual cells and use the host cell to replicate. This causes the cell to burst, allowing new viruses to infect surrounding cells.

Species: The smallest group of classifying organisms, all of which are able to interbreed to produce fertile offspring.

Answers and quick quizzes at **www.hoddereducation.co.uk/myrevisionnotesdownloads**

Measles

Measles is caused by a virus and can be fatal if complications occur. Most young children are vaccinated against measles. Its symptoms include a fever and red skin rash. It is spread through the air in droplets from coughs and sneezes.

HIV/AIDS

Human immunodeficiency virus (HIV) is a virus that is spread by sexual contact during exchange of bodily fluids. This can also occur when blood is swapped in shared needles used by drug users. Initially an infected person will feel flu-like symptoms. Unless an infected person is given antiviral drugs, the virus will attack their body's immune system. HIV disease is called AIDS (acquired immune deficiency syndrome) when the person's immune system can no longer defend them.

Tobacco mosaic virus

The tobacco mosaic virus infects many plants including tomatoes, not just tobacco. You can learn about this in the section on plant disease.

Bacterial diseases

REVISED

Bacteria are alive. They are prokaryotes, so do not have a nucleus. Not all bacteria cause disease. Many, including those in your digestive system, are useful to us. Bacteria live on or in living organisms and are often found in places such as mouths, noses and throats. They do not need the body's individual cells to live, as viruses do. Pathogenic bacteria may produce poisons (toxins) that damage tissues and make us ill.

Salmonella

The *Salmonella* bacterium causes food poisoning. It is often spread in food prepared in unhygienic conditions, or that is undercooked or reheated. Symptoms of infection include fever, abdominal (tummy) cramps, vomiting and diarrhoea. All poultry in the UK are vaccinated against *Salmonella*.

Gonorrhoea

Gonorrhoea is a sexually transmitted disease (STD) spread by sexual contact. It causes a thick yellow or green discharge from the vagina or penis. Urinating is painful. Initially the bacterium that causes the infection was easily treated by penicillin (the first antibiotic). Latterly it has evolved to be resistant to this drug. Treatment is now by more recently discovered antibiotics. Its spread can be reduced by barrier contraception such as a condom.

Fungal diseases

REVISED

Fungi are eukaryotes like animals and plants. Their cells have a nucleus. They have cell walls like plants, but these are made from **chitin** and not cellulose. They can be singled-celled like yeast or multi-celled like mushrooms. Not all fungi cause disease. Some like yeast are very important economically in making bread and beer.

> **Chitin:** A polymer made from sugars that forms the cell walls of fungi and the exoskeleton of insects.

Rose black spot

Black spot is a fungal pathogen that infects roses. You can learn about this in the section on plant disease.

Protist diseases

Protists are also eukaryotes. They are single- or multi-celled but they do not have tissues, organs nor organ systems like animals, plants and fungi.

Malaria

One of the most common diseases caused by a protist is malaria. This protist is spread in blood sucked by mosquitos. These are therefore the vector for the spread of the disease. The protist causes recurring fevers that can be fatal if not treated. Reduction in the spread of malaria occurs by preventing mosquitos from breeding and by using mosquito nets and sprays to avoid being bitten.

Human defence systems

The first line of defence

The first line of defence stops pathogens from entering your body. These defences are non-specific. They are described in Table 3.1.

Table 3.1 Your first line of defence against infection by pathogens.

Type	Description
Skin	Your skin is an organ. It has many functions including insulation and sensitivity but also provides a barrier that almost completely covers you to prevent attack by pathogens. When your skin is broken with a cut or graze, your body works very hard to form a scab to prevent pathogens entering.
	Skin does not cover your mouth or eyes. Here your body produces enzymes called **lysozymes** that attack bacteria by breaking down their cell walls.
Nose	Your nose has hairs and produces mucus to trap pathogens you might have breathed in.
Trachea and bronchi	The cells that line your airways (trachea and bronchi) possess tiny hair-like projections called **cilia**. In between these ciliated cells are goblet cells which produce mucus. This traps any pathogens that have bypassed the hairs and mucus in your nose. The hairs of the ciliated cells beat in a rhythmical motion to waft the mucus and its trapped pathogens up to your throat. When you clear your throat, you swallow this mucus into your stomach where any pathogens are killed.
Stomach	Your stomach contains hydrochloric acid. This does not break down food directly but is strong enough to kill many bacterial pathogens that enter through your mouth or nose.

Lysozymes: Antibacterial enzymes found in your tears to prevent eye infections.

Cilia: Tiny hair-like projections from ciliated cells that waft mucus out of the gas exchange system.

Revision activity

Draw out this table with only the headings along the top and the first column on the left. Try to fill in the rest of the table from memory to help you to revise.

The second line of defence

If a pathogen passes your first line of defence, it is attacked by your second line of defence. This is again non-specific, so all pathogens are attacked in the same way. White blood cells called phagocytes attack all pathogens that have evaded your first line of defence. Their cell membrane flows around the pathogens engulfing them in a vacuole. Enzymes within the vacuole then attack the pathogen cell walls and membranes. This process is called phagocytosis.

Answers and quick quizzes at **www.hoddereducation.co.uk/myrevisionnotesdownloads**

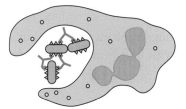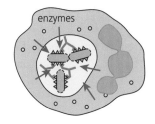
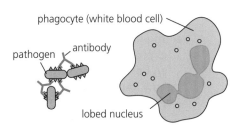

(a) Antibodies cause pathogens to clump together

(b) Phagocyte flows around pathogens to engulf them in a vacuole

(c) Enzymes added to vacuole to break down pathogen cell walls and membranes

Figure 3.1 A white blood cell (phagocyte) engulfing pathogens.

The third line of defence

The third line of defence attacks pathogens in a more specific way. Another type of white blood cells called a lymphocyte produces **antibodies** that specifically match proteins called antigens found on the outside of pathogens. Every pathogen has different antigens and so every time you are infected by a pathogen your lymphocytes produce different antibodies. Your lymphocytes 'remember' each pathogen and can produce more of the specific antibodies faster if you are exposed to the same pathogen again. This means you do not catch the same common cold each winter. There are in fact several hundred different common colds. Antibodies cause pathogens to stick together making it easier for phagocytes to engulf and destroy them.

Your lymphocytes also produce **antitoxins**. These are a special type of antibody that can neutralise the toxins produced by some pathogens that make you feel ill.

> **Exam tip**
>
> You should be able to describe the non-specific defence systems and the role of the immune system in the defence against disease.

> **Antibodies**: Proteins produced by lymphocytes that recognise the antigen of pathogens and help to clump them together.
>
> **Antitoxin**: A protein produced by your body to neutralise harmful toxins produced by pathogens.

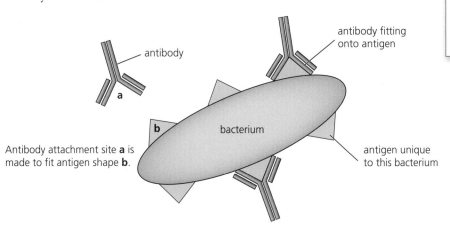

Antibody attachment site **a** is made to fit antigen shape **b**.

Figure 3.2 How an antibody fits onto the antigen of a pathogen.

Vaccination

REVISED

Vaccinations protect us from disease. If a large proportion of a population is vaccinated it is much less likely that a disease will spread. This is called herd immunity. A vaccination is a small quantity of dead or inactive form of a pathogen. This is introduced into the body, often by injection. This stimulates the lymphocytes (white blood cells) to produce antibodies.

Shortly after a vaccination you may feel a little sick. This is your body fighting the disease and is called the initial exposure. However, if you were to encounter a more severe case of the pathogen in later life (a secondary exposure), then your lymphocytes 'remember' the infection and produce more antibodies faster. In doing so, you are much less likely to fall ill.

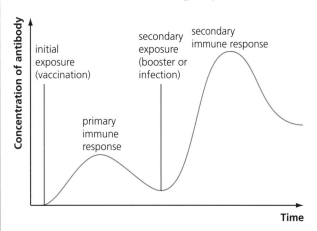

Figure 3.3 This graph shows the rate at which antibodies are produced after the first and second exposures to a pathogen.

Antibiotics and painkillers

REVISED

Antibiotics

Antibiotics, such as penicillin, are medicines that cure bacterial disease by killing bacteria within an infected person's body. They do not kill viral pathogens, so you will never be prescribed antibiotics for a common cold (caused by a virus). It is difficult to develop drugs that kill viruses within cells without hurting the cells themselves.

Since the discovery of the first antibiotic by Sir Alexander Fleming (1881–1955) around a hundred years ago, millions of lives have been saved by their use. However, some bacteria are rapidly evolving resistance to our antibiotics. In fact, they are doing this much faster than we can find new ones. Doctors are worried that soon many of our antibiotics will not work anymore and that deaths will increase again.

Painkillers

Painkillers are drugs that reduce or stop pain. Some painkillers were discovered in plants. Aspirin naturally occurs in the bark of willow trees. Others like paracetamol are artificial and have been designed by chemists. Painkillers are used to reduce the symptoms of disease such as pain, swelling and fever. They do not kill the pathogens themselves.

Discovery and development of drugs

REVISED

Drugs are chemicals that have a biological effect on the organism that takes it. Drugs can be natural or man-made. They can be helpful like medicines or harmful like many illegal drugs. Some are more easily obtained like alcohol and tobacco. The use of others is controlled by doctors and can only be taken with a medical prescription.

Traditionally drugs were extracted from plants and micro-organisms. The heart drug digitalis originated from foxglove plants. The painkiller aspirin originated in the bark of willow trees. The antibiotic penicillin was discovered by mistake from the *Penicillium* mould. Most new drugs are developed by chemists in the pharmaceutical industry, although the starting point may be a natural chemical compound.

Modern drug development

Drug development is a long and expensive process that takes many years and costs hundreds of millions of pounds. Only around 0.1% of drugs pass this testing process. The process is shown in Table 3.2.

Table 3.2 The stages of modern drug development

Stage	Description
One	This uses **computer modelling** to look at the structure of the drug and the interactions it might have on the human body.
Two	Stage two involves laboratory tests. These can be on cells grown in the laboratory or on animals. The results from these are used to predict how the drug will affect humans.
Three	The final stage involves human trials. In the first part of this stage, the drug is given to a small number of healthy volunteers. This is to determine the correct dosage. The second round of tests here are given to sick patients to see how effective it is (its **efficacy**). The final stage involves tests on much larger numbers of volunteers to check dosage and efficacy. In some of the tests, some of the volunteers are given **placebo** doses. These look the same as the active drugs. They are used to eliminate the placebo effect, which is where people can feel better because they think they have taken the drug even if they actually have not. Some of the tests are called '**double blind**'. In these, neither the doctors or the volunteers know if they have been give the drug or the placebo. This eliminates any bias.

Now test yourself

TESTED

1 What are pathogens?
2 What are vectors? Give an example of a vector in your answer.
3 State an example of a communicable disease caused by a virus, bacterium and fungus.
4 What do all examples of the first line of defence stop?
5 What are lysozymes?
6 How are phagocytes and lymphocytes different?
7 What do lymphocytes produce to neutralise harmful toxins produced by pathogens?
8 What is herd immunity and why is it important?
9 What type of pathogen are antibiotics useless to treat?
10 Describe a double-blind trial.

Answers online

Summary

- Communicable diseases are those which are transmitted from person to person. Pathogens are viruses, bacteria, protists and fungi that cause communicable diseases to plants and animals. They are transmitted by direct contact, water or air.
- Bacteria reproduce inside the body and can produce toxins. Viruses reproduce inside cells.

- Measles is caused by a virus. Symptoms include a fever and red skin rash. HIV is a virus which initially causes flu-like symptoms. HIV develops into AIDS when a person's immune system cannot function anymore. Tobacco mosaic virus infects plants causing discoloured leaves which reduces growth.

- *Salmonella* food poisoning is spread in food prepared in unhygienic conditions. The symptoms include fever, abdominal cramps, vomiting and diarrhoea. Gonorrhoea is a bacterial STD which causes thick yellow or green discharge from the vagina or penis.
- Rose black spot is caused by a fungus. Infected plants have discoloured leaves which reduces growth. Malaria is transmitted by a protist carried within a mosquito. Symptoms include recurring fever and can be fatal.
- Non-specific defence systems against infection include skin, hairs within the nose, ciliated cells in the trachea and bronchi and stomach acid.
- If an invading pathogen passes these defences, the immune system attacks. White blood cells called phagocytes engulf and destroy pathogens. White blood cells called lymphocytes produce antibodies which 'clump' together pathogens.

They also produce antitoxins to neutralise toxins produced by the pathogen. Lymphocytes 'remember' pathogens and are produced more quickly in subsequent exposures to them.
- Antibiotics are medicines which kill bacteria but not viruses. Their use has saved many lives. However, bacteria like methicillin-resistant *Staphylococcus aureus* (MRSA) are evolving to be antibiotic resistant. This is of great concern. Painkillers are used to treat symptoms but not kill pathogens.
- Traditionally drugs were extracted from plants and micro-organisms. Newer drugs are often synthesised by chemists. The discovery and development of new drugs is expensive and time-consuming. This tests toxicity, efficacy and dosage. Preclinical trials are completed on cells, tissues and live animals. Clinical trials then use healthy volunteers and patients.

Exam practice

1 What causes rose black spot? [1]
   A Viruses
   B Fungi
   C Bacteria
   D Protists
2 Who discovered the first antibiotic? [1]
   A Louis Pasteur
   B Charles Darwin
   C Alexander Fleming
   D Alfred Russel Wallace
3 Define the term antiseptic. [1]
4 What are the symptoms of the tobacco mosaic virus and what effect does this have upon the plant? [2]
5 What part of a pathogen do antibodies bind with? [1]
6 Label the two points on the graph shown by the lines. Describe the shape of this graph. [4]

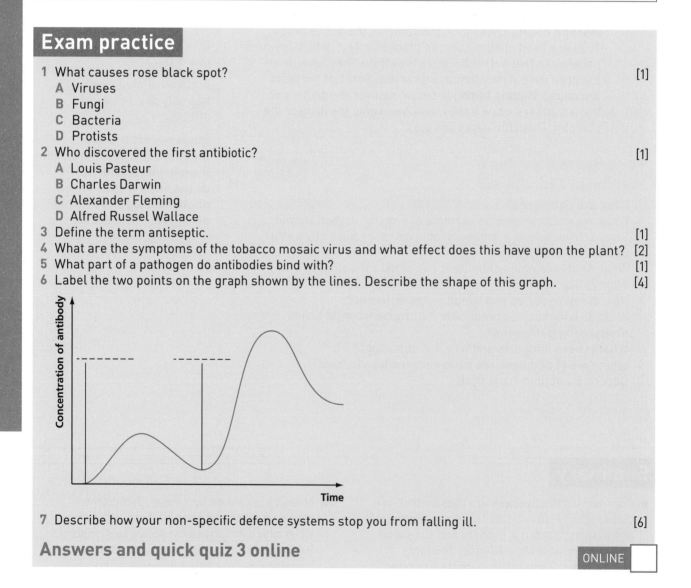

7 Describe how your non-specific defence systems stop you from falling ill. [6]

Answers and quick quiz 3 online

ONLINE

4 Bioenergetics

Photosynthesis

Photosynthetic reaction

REVISED

Plants take in water through their roots and carbon dioxide through their leaves. These reactants are converted into glucose during photosynthesis. Oxygen is made as a by-product of this reaction. The word equation for this reaction is:

$$\text{carbon dioxide} + \text{water} \xrightarrow{\text{light in}} \text{glucose} + \text{oxygen}$$

The balanced symbol equation for photosynthesis is:

$$6CO_2 + 6H_2O \xrightarrow{\text{light in}} C_6H_{12}O_6 + 6O_2$$

Photosynthesis only occurs in light. It is an **endothermic reaction** so requires energy from light. Photosynthesis occurs in sub-cellular structures called chloroplasts, which contain a green compound called chlorophyll. Chlorophyll is essential for photosynthesis. Chloroplasts are found in high numbers in palisade mesophyll cells which are in turn found in the top layers of leaves.

> **Exam tip**
>
> In both the word and symbol equations, the two reactants and the two products can be either way around.

> **Endothermic reaction**: A reaction that requires heat to be absorbed to work.

Photosynthetic algae

Algae can photosynthesise as well as plants. Some algae are single-celled whilst others like seaweed are larger. Some algae have chlorophyll like plants, but others use different coloured photosynthetic pigments. These are seen in seaweeds which are green, brown and red. More than two-thirds of the oxygen made every day is by photosynthetic algae in our oceans and not plants on land.

> **Typical mistake**
>
> Plants complete photosynthesis to make glucose. Oxygen is a by-product. They do not make it for us.

Rate of photosynthesis

REVISED

The rate of reaction is how quickly it occurs. The rate at which plants and algae photosynthesise decreases when:
- temperatures fall (less kinetic energy present)
- carbon dioxide levels drop
- light intensities reduce (see below)
- plants do not have sufficient chlorophyll.

H If one of more of these conditions occurs, the rate of photosynthesis becomes limited. This or these are then '**limiting factors**'. It is important for farmers to reduce any limiting factors to ensure the maximum **yield** of their crops. They do this by:
- keeping plants warm in greenhouses or polytunnels
- keeping burners in greenhouses or polytunnels to produce carbon dioxide
- providing plants with maximum light levels.

> **Limiting factor**: Anything that reduces or stops the rate of a reaction.
>
> **Yield**: The amount of an agricultural product.

Required practical 5

Light intensity and photosynthesis

Aim: To investigate the effect of light intensity on the rate of photosynthesis using an aquatic organism such as pondweed

Equipment: Lamp, metre rulers, boiling tube, pondweed, water.

Method:
1 Set up the equipment as shown in Figure 4.1.
2 Place the boiling tube with pondweed 10 cm from the lamp.
3 Switch the light on and wait 2 minutes for the plant to acclimatise.
4 Record the number of bubbles of oxygen given off in 1 minute.
5 Move the boiling tube a further 10 cm from the lamp and repeat.
6 Repeat by increasing the distance every 10 cm up to 60 cm.

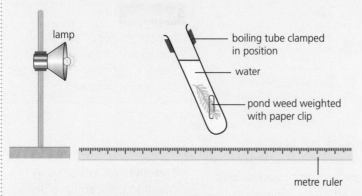

Figure 4.1 The equipment used to investigate the effects of light intensity on the rate of photosynthesis.

Results: Your results will look like those in Figure 4.2.

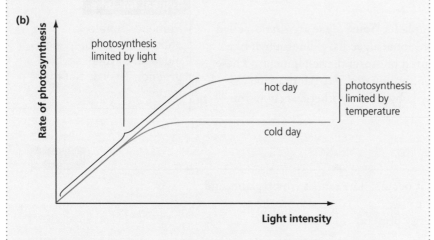

Figure 4.2 The effect of increasing light intensity on the rate of photosynthesis.

Uses of glucose from photosynthesis

REVISED

Photosynthesis provides plants and algae with chemical energy in the form of glucose. They use this in five key ways shown in Figure 4.3.

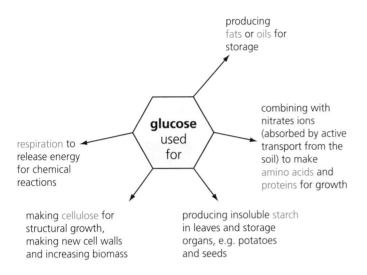

producing fats or oils for storage

glucose used for

combining with nitrates ions (absorbed by active transport from the soil) to make amino acids and proteins for growth

respiration to release energy for chemical reactions

making cellulose for structural growth, making new cell walls and increasing biomass

producing insoluble starch in leaves and storage organs, e.g. potatoes and seeds

Figure 4.3 How plants make use of the glucose produced in photosynthesis.

Plants or algae doing photosynthesis are present at the bottoms of almost all food chains on our planet. So, photosynthesis provides the energy that supports almost all life on Earth including your own. Without photosynthesis there would be very little life on our planet.

Now test yourself

TESTED

1 State the word equation for photosynthesis.
2 State the balanced symbol equation for photosynthesis.
3 Besides plants, what other group of organisms completes photosynthesis?
4 Why is photosynthesis an endothermic reaction?
5 In what type of cells does most photosynthesis in leaves occur?
6 What four factors affect the rate of photosynthesis?
H▶ 7 How do farmers reduce limiting factors for photosynthesis?
8 Define the term yield.
9 State the equipment that you would use to investigate how light intensity affects the rate of photosynthesis.
10 Where are photosynthetic organisms usually found in a food chain?

Answers online

Respiration

Aerobic respiration

REVISED

Energy is released from glucose during **aerobic** respiration. This reaction is essential for many living organisms. It occurs continuously in specially adapted cell components called mitochondria which are found in the cytoplasm. The energy that is released is used to complete the seven life processes: movement, reproduction, respiration, sensitivity, nutrition, excretion and growth. The word equation for this reaction is:

glucose + oxygen $\xrightarrow{\text{energy out}}$ carbon dioxide + water

The balanced symbol equation for respiration is:

$C_6H_{12}O_6 + 6O_2 \xrightarrow{\text{energy out}} 6CO_2 + 6H_2O$

Aerobic: In the presence of oxygen.

Exam tip

In both the word and symbol equations, the two reactants and the two products can be either way around.

Respiration is an **exothermic reaction**. It transfers energy into its surroundings. Unlike photosynthesis, respiration occurs at all times not just during the day.

> **Exothermic reaction**: A reaction that gives out heat energy.

Conversion of energy in respiration (and photosynthesis)

REVISED

When the equations for photosynthesis and respiration are looked at, they initially seem to be the opposite of each other.

> **Exam tip**
>
> You should also be able to describe cellular respiration as an exothermic reaction which is continuously occurring in all cells.

The reactants in respiration are the same as the products in photosynthesis and the other way around. However, there is one crucial difference. Photosynthesis is endothermic and so takes energy from its surroundings (light) and stores this in glucose. Respiration is exothermic and so releases energy from glucose.

The equations for photosynthesis and respiration work together for almost all life on our planet. The Sun's energy drives almost all food chains. Energy transferred by light (mainly from the Sun) is converted by plants and algae into a chemical store of energy called glucose during photosynthesis. Energy for life processes is then transferred from glucose during respiration. This energy has two main uses. It is converted into:

1 Heat energy to keep **homoeothermic** (warm-blooded) animals warm.
2 A chemical store of energy that is used in reactions to build larger molecules and processes like movement.

> **Homoeothermic**: Warm blooded, like birds and mammals.

Anaerobic respiration

REVISED

Anaerobic respiration occurs when there is not enough oxygen to complete anaerobic respiration. Times like this occur when you have been exercising vigorously. You are unable to breathe quickly enough or deep enough to supply your cells with sufficient oxygen. This is the equation for anaerobic respiration:

$$\text{glucose} \xrightarrow{\text{energy out (only 5\%)}} \text{lactic acid}$$

When your cells respire anaerobically, there is insufficient oxygen to make carbon dioxide and water. So, an intermediary **product** called lactic acid is made instead. Many scientists think that lactic acid build-up in muscles causes cramp. During long periods of vigorous exercise, your muscles become fatigued and stop contracting efficiently.

> **Anaerobic**: In the absence of oxygen.
>
> **Product**: The substance or substances produced in a reaction.
>
> **Oxygen debt**: The temporary shortage of oxygen in respiring tissues and organs.

Crucially, only 5% of the energy released in aerobic respiration is transferred during anaerobic respiration. The remaining 95% is stored within the lactic acid.

When you have finished exercising, your breathing rate and volume (depth) do not reduce to normal immediately. They remain high until your body has absorbed enough oxygen to pay your body back the oxygen you owe it. This is paying your **oxygen debt**. When this happens, the lactic acid can be fully broken down into carbon dioxide and water. This releases the remaining 95% of the energy that was originally stored in glucose. This process is shown in the following equation:

> **Exam tip**
>
> You should also be able to compare aerobic and anaerobic respiration.

$$\text{lactic acid} + \text{oxygen} \xrightarrow{\text{energy out (only 95\%)}} \text{carbon dioxide} + \text{water}$$

H Lactic builds up in cells that are respiring anaerobically. This diffuses into the bloodstream from high concentration in the cells to a lower concentration in the blood. This is transported to the liver. Here it diffuses from a high concentration in the blood to a lower concentration in the liver. It is then converted back to glucose.

Anaerobic respiration in plants and micro-organisms

All cells in your body must complete respiration to release the energy they need to live. The same is true of all plants and micro-organisms as well. Some micro-organisms such as the yeast fungus respire anaerobically too:

$$\text{glucose} \xrightarrow{\text{energy out}} \text{ethanol} + \text{carbon dioxide}$$

This reaction in yeast is called **fermentation**. This is an economically important reaction for we use it to make bread and beer. Ethanol is commonly called alcohol. It is present in about 4% in beer, 12% in wine and 40% in spirits. Bread is not alcoholic because it is baked before we eat it. This evaporates away any alcohol.

> **Fermentation**: The chemical break-down of glucose into ethanol and carbon dioxide by respiring micro-organisms such as yeast.

Metabolism

Metabolism is the sum of all the chemical reactions in your body. These include the digestion of food, aerobic and anaerobic respiration and protein synthesis. Your metabolism is regulated by your thyroid gland. These chemical reactions either break down large molecules into smaller ones or the reverse.

Break-down reactions

Break-down reactions turn larger, often more complicated molecules into smaller ones. These reactions do not often require much energy from respiration. Any excess proteins you obtain from your diet are broken down to form amino acids in your liver. These in turn are broken down by removal of an amine group. This forms urea which is excreted by your kidneys in your urine.

Synthesis reactions

Synthesis reactions turn smaller, often less complicated molecules into larger ones. These reactions often do require energy from respiration. The following are examples of synthesis reactions:

- Glucose is produced by plants and algae during photosynthesis. They can store the glucose as insoluble starch or convert it to cellulose to make cell walls. In animals, excess glucose can be stored as **glycogen** which is stored in the liver.
- Lipids are fats (solids) and oils (liquids). These are made from three molecules of fatty acid and one of glycerol. Lipids are needed for cell membranes and as a store of chemical energy (fat).
- Glucose and nitrate ions are needed to make amino acids. These are joined together in the correct sequence to make proteins during protein synthesis.

> **Exam tip**
>
> You should also be able to explain the importance of sugars, amino acids, fatty acids and glycerol in the synthesis and break-down of carbohydrates, proteins and lipids.

> **Glycogen**: An insoluble store of glucose in the liver.

Now test yourself

11 State the word equation for respiration.
12 State the symbol equation for respiration.
13 Why is respiration an exothermic reaction?
14 When do plants complete respiration and photosynthesis?
15 What are the two main uses of energy produced in respiration in warm-blooded animals?
16 What does anaerobic mean?
17 State the word equation for anaerobic respiration.
H▶ 18 What is lactic acid broken down into?
19 State the word equation for fermentation.
20 What two types of reactions make up metabolism?

Answers online

Summary

- The word equation for photosynthesis is:

$$\text{carbon dioxide + water} \xrightarrow{\text{light in}} \text{glucose + oxygen}$$

- The symbol equation for photosynthesis is:

$$6CO_2 + 6H_2O \xrightarrow{\text{light in}} C_6H_{12}O_6 + 6O_2$$

- Photosynthesis is an endothermic reaction in which energy is transferred from the environment to the chloroplasts in plant cells by light.
- The rate of photosynthesis is affected by temperature, light intensity, carbon dioxide concentration and the amount of chlorophyll.
- These limiting factors can interact. They are important in the economics of obtaining optimum conditions in greenhouses for growing food.
- Glucose produced in photosynthesis is used for respiration, converted into insoluble starch for storage, used to produce fats or oils for storage, used to produce cellulose for growth and amino acids for protein synthesis.
- Respiration can occur in the presence of oxygen (aerobic) or absence (anaerobic).
- The energy transferred by respiration supplies all the energy needed for living organisms. This is needed for chemical reactions to build larger molecules, movement and keeping warm (in warm-blooded animals).
- The equation for aerobic respiration is:

$$\text{glucose + oxygen} \xrightarrow{\text{energy out}} \text{carbon dioxide + water}$$

- The balanced symbol equation for aerobic respiration is:

$$C_6H_{12}O_6 + 6O_2 \xrightarrow{\text{energy out}} 6CO_2 + 6H_2O$$

- During exercise, the heart and breathing rates increase together with the breath depth. This supplies respiring cells with more oxygenated blood. Without sufficient oxygen, anaerobic respiration occurs.
- The equation for anaerobic respiration is:

$$\text{glucose} \xrightarrow{\text{energy out (only 5\%)}} \text{lactic acid}$$

- Only around 5% of the energy is released in anaerobic respiration. The rest remains within lactic acid. Oxygen debt is the amount of extra oxygen that is needed to react with and remove lactic acid. When an oxygen debt has been paid, lactic acid is broken down into carbon dioxide and water and the rest of the energy is released.
H - Lactic acid is transported to the liver where it is converted back to glucose.
- The equation for anaerobic respiration in plant and yeast cells is:

$$\text{glucose} \xrightarrow{\text{energy out}} \text{ethanol + carbon dioxide}$$

- Anaerobic respiration in yeast is called fermentation. This is an economically important reaction in the making of bread and alcoholic drinks.
- Metabolism is the sum of all the reactions in a cell or organism. It includes converting glucose to starch, glycogen and cellulose. It includes forming lipids from fatty acids and glycerol. It includes using glucose and nitrate ions to form amino acids to make proteins. It also includes respiration and break-down of urea.

Exam practice

1 In what cell component does photosynthesis occur? [1]
   A Chlorophyll
   B Ribosomes
   C Mitochondria
   D Chloroplasts
2 What is the product of anaerobic respiration (fermentation) in micro-organisms? [1]
   A Lactic acid
   B Ethanol and carbon dioxide
   C Water and carbon dioxide
   D Glucose and oxygen
3 Define the term metabolism. [1]
4 State what happens to excess amino acids in your blood. What type of reaction is this? [2]
5 Why is photosynthesis important for all life on Earth not just plants and algae. [2]
H▶ 6 State the word (standard tier) or symbol equations for photosynthesis. [2]
7 Describe the method you would use to investigate the effects of light intensity on photosynthesis. [4]
8 Describe the uses of glucose from photosynthesis. [6]
9 Compare and contrast the reactions of aerobic respiration and photosynthesis. [6]

Answers and quick quiz 4 online

ONLINE

5 Homeostasis and response

The human nervous system

Homeostasis

REVISED

The millions of cells in your body have optimum conditions. They require glucose for respiration, water and a sufficient temperature to function. **Homeostasis** is the maintenance of these and many other conditions. We call it the maintenance of a constant internal environment. These changes happen without you knowing. They are automatic or involuntary.

> **Homeostasis**: The maintenance of a constant internal environment.

> **Exam tip**
>
> You should also be able to describe homeostasis as the regulation of the internal conditions of a cell or organism to maintain optimum conditions.

Structure and function of the nervous system

REVISED

Your nervous system controls your voluntary and involuntary actions. It allows you to react to your surroundings and coordinate your behaviour. It is made from millions of nerve cells which transmit and receive millions of messages each day. Your nervous system is made up from:

- Your **central nervous system (CNS)** – your brain and spinal cord
- Your peripheral nervous system – the millions of nerves that criss-cross the rest of your body.

Nerves are made from bundles of individual neurones.

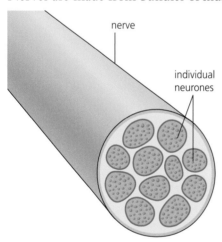

Figure 5.1 A bundle of neurones is a nerve.

> **Exam tip**
>
> You should also be able to explain how the structure of the nervous system is adapted to its functions.

> **Central nervous system (CNS)**: The brain and spinal cord.
>
> **Receptor**: A cell or group of cells at the beginning of a pathway of neurones that detects a change and generates an electrical impulse.

Sending electrical impulses

All messages sent along neurones in your nervous system are electrical. These messages travel very quickly. The electrical signal is generated by a specific type of cell called a **receptor**. You have these cells in all your sense organs including your skin. Some areas of your skin like your lips and finger tips are very sensitive. Here there are more receptors than in less sensitive areas like your elbow.

Table 5.1 Your senses, the organs and stimuli involved.

Sense	Organ	Stimuli
Sight	Eyes	Light
Hearing	Ears	Sound
Taste	Tongue	Chemicals in food
Smell	Nose	Chemicals in air
Touch	Skin	Touch, pressure, temperature, pain and itch

Revision activity

Draw out this table with only the headings along the top and the first column on the left. Try to fill in the rest of the table from memory to help you to revise.

Sensory, relay and motor neurones

There are three main types of neurone. Those that carry signals from receptors towards your central nervous system are called **sensory neurones**. Those that carry signals around your brain and spinal cord are called **relay neurones**. Those that carry signals away from your central nervous system are called **motor neurones**.

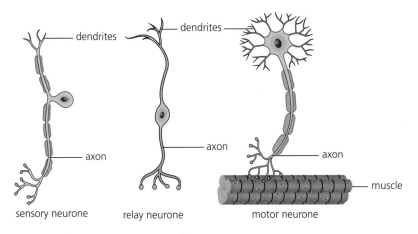

Figure 5.2 The three types of neurone.

Motor neurones end in muscles or glands. Muscles contract or relax to move parts of your body. Glands release hormones into the blood. We call both muscles and glands **effectors**. The pathway of an electrical signal from receptor to effector is:

stimulus → receptor → sensory neurones → relay neurones → motor neurones → effector → response

Collectively, the three types of neurone are called coordinators.

Synapses

There is more than one nerve that links each of your fingers to your brain. These nerves form a network. If one nerve is damaged, the signal can be rerouted and still reach your brain. If each electrical signal passes along multiple nerves, there must be gaps between them. These are called **synapses**.

When an electrical signal reaches the end of a neurone, it is quickly converted into a chemical one to cross the synapse. The ends of neurones are branched. At the tips of these branches, are special areas that make **neurotransmitters**. These chemicals diffuse across the synapse, bind to receptors and restart the electrical signal in the **dendrites** of the next nerve. This happens extremely quickly.

Sensory neurone: A neurone that carries an electrical impulse from a receptor towards the central nervous system.

Relay neurone: A neurone that carries an electrical impulse within the central nervous system.

Motor neurone: A neurone that carries an electrical impulse away from the central nervous system to an effector (muscle or gland).

Effector: A muscle or a gland.

Synapse: A gap between the axon of one nerve and the dendrites of another where neurotransmitters transmit the impulse.

Neurotransmitter: A chemical substance released at the end of one neurone that diffuses across a synapse to begin a second electrical impulse in another neurone.

Dendrites: The branched beginnings of neurones, which can detect neurotransmitters and start another electrical impulse.

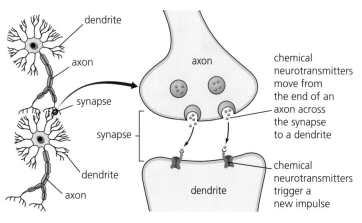

Figure 5.3 How signals move from one nerve cell to another across a synapse.

labels in figure: dendrite, axon, synapse, synapse, dendrite, axon, axon, chemical neurotransmitters move from the end of an axon across the synapse to a dendrite, chemical neurotransmitters trigger a new impulse, dendrite

Reflex arc: The movement of an electrical impulse that avoids the brain to save time and so prevent damage to your body.

Reflex response: An automatic response that you do not think about.

Concordant results: Results that are similar.

The reflex arc

Some of your reactions are automatic. You do not think about moving your hand if you put it on a hot radiator. You just move it. These are called **reflex responses**. Here the electrical signal is not initially transferred to the conscious region of your brain. The signal is started by receptors and travels along sensory neurones (like normal) to your spinal cord. Here relay neurones immediately send a signal down motor neurones to your muscles, which contract and move your body. The signal is sent to your conscious brain shortly afterwards. This makes your reactions quicker which potentially reduces damage to your body.

Exam tip

You should be able to explain how the structure of synapses, and sensory, relay and motor neurones all relate to their function. You should also be able to explain the importance of reflex actions.

Required practical 6

Reaction times

Aim: To plan and carry out an investigation into the effect of a factor on human reaction time.

Equipment: Ruler

Method:
You will need a partner to complete this investigation.
1 Person A should hold the end of a ruler between the finger and thumb of person B. Without warning, person A should drop the ruler for person B to catch.
2 The shorter the distance the ruler drops, the faster the reactions of person B.
3 Charts that convert distance to reaction time are easily found on the internet.
4 Repeat this until you have three **concordant results**, ignore anomalous results and calculate the mean.
5 Drink a can of caffeinated drink and wait a short period for it to be absorbed into your bloodstream.
6 Repeat and compare your results.

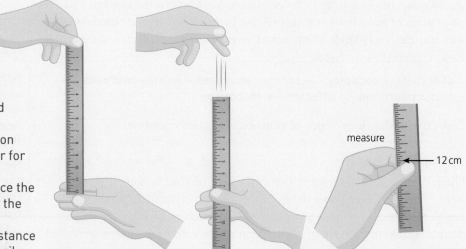

Figure 5.4 How to test your partner's reaction time.

measure — 12 cm

Results: Most people find that their reaction time reduces after drinking a caffeinated drink. This means their reactions are faster.

Now test yourself

TESTED ☐

1 Define the term homeostasis.
2 What are effectors?
3 Describe the pathway of an electrical signal from receptor to effector.
4 How is the pathway of a reflex action different from a 'normal' nervous response?

Answers online

Hormonal control in humans

Human endocrine system

REVISED ☐

Your **endocrine system** is a group of glands that secrete hormones into your blood. These glands, their hormones and functions are given in Table 5.3 and shown in Figure 5.5. Hormones travel in the blood and so hormonal responses are usually slower than electrical signals that are carried by the nerves of the nervous system.

> **Endocrine system**: The system of glands that secrete hormones into the circulatory system.

Table 5.3 **Common examples of hormones and their functions.**

Hormone	Produced	Target organ	Function
ADH (anti-diuretic hormone)	Pituitary gland	Kidney	Controls the concentration of water in urine
TSH (thyroid-stimulating hormone)	Pituitary gland	Thyroid	Controls the release of hormones from your thyroid gland
Adrenaline	Adrenal gland	Heart (and other vital organs)	Prepares the body to fight or run away (flight)
Insulin and glucagon	Pancreas	Liver	Insulin increases and glucagon decreases the conversion of blood glucose to glycogen
Thyroid hormones (e.g. thyroxine)	Thyroid	Various	Control how quickly you use energy, make proteins and how sensitive your organs are to other hormones
Oestrogen	Ovaries	Reproductive organs	Controls puberty and the menstrual cycle in women
Testosterone	Testes	Reproductive organs	Controls puberty in men

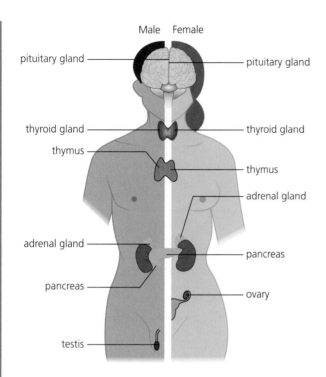

Male Female

pituitary gland —— —— pituitary gland

thyroid gland —— —— thyroid gland

thymus —— —— thymus

—— adrenal gland

adrenal gland —— —— pancreas

pancreas —— —— ovary

testis ——

Figure 5.5 The positions of key glands in your endocrine system.

The **pituitary gland** in your brain is your 'master gland'. It secretes hormones that directly control growth and blood pressure, and partly controls how your kidneys, and ovaries or testes function. It also partly controls pregnancy and childbirth. It does this by releasing hormones that control other glands.

Pituitary gland: A gland in your brain that produces growth hormones, anti-diuretic hormone (ADH), thyroid-stimulating hormone (TSH), follicle-stimulating hormone (FSH) (in women) and luteinising hormone (LH) (again in women).

Exam tip

You should also be able to describe the principles of hormonal coordination and control by the endocrine system. You should also be able to identify the position of the glands in Figure 5.5.

Control of blood glucose concentration

REVISED

Your digestive system breaks down food into smaller, soluble molecules that are absorbed into your blood. Carbohydrase enzymes break down carbohydrates like starch into sugars like glucose. Many people eat three times a day but all your cells need glucose all the time to release energy during **respiration**. Your body absorbs excess sugar and stores it for times when you need it. This is an example of homeostasis. The concentration of your blood glucose is monitored and controlled by your pancreas.

Respiration: The release of energy from glucose.

Insulin: A hormone produced in your pancreas that lowers blood glucose by converting it to glycogen and storing it in the liver.

Glucagon: A hormone produced in the pancreas that raises blood glucose by breaking down glycogen stored in the liver.

Insulin and glucagon

When your blood glucose level is too high, your pancreas releases the hormone **insulin**. This travels in the bloodstream to your liver and muscle cells which convert excess glucose into insoluble glycogen. This reduces your blood glucose levels and returns them to normal.

H When your blood glucose level is too low, your pancreas releases the hormone **glucagon**. This travels in the bloodstream to the liver which converts glycogen back to soluble glucose. This is released into your blood so it increases your blood glucose levels and returns them to normal.

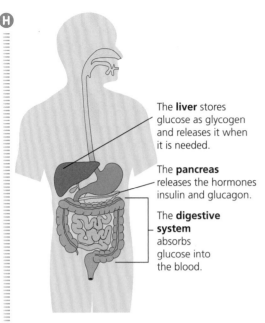

The **liver** stores glucose as glycogen and releases it when it is needed.

The **pancreas** releases the hormones insulin and glucagon.

The **digestive system** absorbs glucose into the blood.

Figure 5.6 Absorption of glucose, control of blood glucose concentration and storage of glucose in the body.

This is an example of **negative feedback control**. This occurs when your body detects a change (too little or too much blood glucose), makes a change and returns itself to normal. This change is therefore an example of homeostasis.

Diabetes

The causes of **type 1 diabetes** are unknown. It develops in children or young adults. It results in a person's immune system mistakenly destroying the insulin-producing cells in their pancreas. This means they cannot produce insulin. There is no cure. People with type 1 diabetes usually inject insulin, eat carefully and exercise regularly to control their blood glucose concentration.

Type 2 diabetes usually develops later in life than type 1. People with type 2 diabetes cannot produce enough insulin or if they can their liver and muscle cells do not respond to it. They do not absorb excess glucose and store it as glycogen. People with type 2 diabetes often feel thirsty, urinate more frequently and feel tired. Type 2 diabetes develops more commonly in people who do not exercise regularly, have a high sugar diet and are obese. There is no cure for type 2 diabetes either. Insulin is not injected (because their cells do not respond to it) but people with type 2 diabetes eat carefully and exercise regularly.

Hormones in human reproduction   REVISED

The sex hormone in men is **testosterone**, which is secreted from the testes. In women, it is the hormones **oestrogen** and **progesterone,** which are produced by the ovaries. These hormones are responsible for puberty in both sexes. Oestrogen and progesterone also regulate the menstrual cycle.

Exam tip

You should also be able to explain how insulin controls blood glucose levels.

Typical mistake

It is important you do not mistake the hormone glucagon for the storage molecule glycogen.

Exam tip

You should also be able to explain how glucagon interacts with insulin in a negative feedback cycle to control blood glucose levels.

Negative feedback control: A homeostatic mechanism by which the body detects a change and makes an adjustment to return itself to normal.

Type 1 diabetes: A medical condition that usually develops in younger people, preventing the production of insulin.

Type 2 diabetes: A medical condition that usually develops in later life, preventing the absorption of insulin.

Exam tip

You should also be able to compare type 1 and type 2 diabetes and explain how they are treated. You should be able to explain the differences between graphs of blood glucose levels of people with and without diabetes.

Testosterone: A male sex hormone produced in the testes that controls puberty.

Oestrogen: A female sex hormone produced in the ovaries that controls puberty and prepares the uterus for pregnancy.

Progesterone: A female sex hormone produced in the corpus luteum that prepares the uterus for pregnancy.

The menstrual cycle

From puberty to the **menopause**, at between 45 and 55 years old, women undergo a 28-day reproductive cycle called the menstrual cycle. Unless a woman is pregnant, the menstrual cycle begins on day one with the break-down of the lining of the uterus from the previous cycle. This process is called **menstruation** or having a period and lasts for a few days. It can be painful and cause cramps. Shortly after this process, the hormone oestrogen is produced by the ovaries and causes the lining of the uterus to start to thicken again. It is preparing itself for a fertilised ovum (egg) to settle and for the woman to become pregnant.

Around day 14 of the cycle, an ovum is released from an ovary. This is called ovulation. In the days following ovulation a woman is at her most fertile. If sperm are ejaculated into her vagina, they can swim upwards past the cervix and through the uterus to meet the ovum in a fallopian tube. If the sperm fertilises the ovum, it is likely to settle into the lining of the uterus and develop into a baby. If this happens, levels of the hormone progesterone stay high in the woman's body. This stops her from having another period, which would result in a natural abortion. If the ovum is not fertilised, then progesterone levels drop towards the end of the cycle and the woman menstruates again.

There are four hormones that work together in the menstrual cycle. They are described in Table 5.4 and their effects are shown in Figure 5.7.

Table 5.4 **The main hormones involved in the control of the menstrual cycle.**

Hormone	Released by	Target organ and effect
Follicle-stimulating hormone (FSH)	Pituitary gland	Ovary ● Causes an ovum to mature in the ovary inside the **follicle** ● Stimulates ovaries to produce oestrogen
Oestrogen	Ovaries	Uterus ● Causes lining to thicken in first half of the cycle ● High oestrogen concentration switches off the release of FSH and switches on the release of LH
Luteinising hormone (LH)	Pituitary gland	Ovary ● Stimulates ovulation (release of the ovum from the ovary)
Progesterone (produced if fertilised ovum implants in uterus)	Ovaries (**corpus luteum**)	Uterus ● Maintains thick uterus lining if fertilised ovum implants ● High concentrations of progesterone in pregnancy stop the cycle

Menopause: The point in a woman's life, usually between 45 and 55, when she stops menstruating after which she cannot become pregnant.

Menstruation: Having a period, as a part of the menstrual cycle.

Follicle-stimulating hormone (FSH): A hormone produced by the pituitary gland that causes an ovum to mature in an ovary and stimulates the ovary to produce oestrogen.

Luteinising hormone (LH): A hormone produced by the pituitary gland that stimulates ovulation.

Corpus luteum: After ovulation the empty follicle turns into this and releases progesterone.

Follicle: A structure in an ovary in which an ovum matures.

Answers and quick quizzes at **www.hoddereducation.co.uk/myrevisionnotesdownloads**

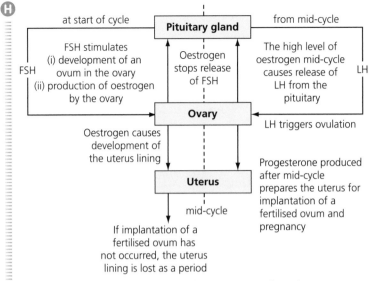

Figure 5.7 **Hormones control the menstrual cycle.**

Exam tip

You should also be able to describe the roles of hormones in human reproduction including those in the menstrual cycle.

Contraception

REVISED

Contraception is a name given to devices or methods that stop women becoming pregnant. It is an ethical issue because some people disagree with it for religious or moral reasons. The Catholic Church wants its followers to practice family planning or abstinence and not use other forms of contraception. Family planning involves the man not ejaculating inside the woman's vagina at the time when an ovum is likely to be in the fallopian tubes (approximately days 13 to 17 of the menstrual cycle). Abstinence is stopping having sex altogether.

A widespread form of contraception is the condom. The condom is an example of barrier contraception because it covers the erect penis and stops sperm from reaching the ovum. The diaphragm is another form of barrier contraception. It is a small plastic dome that sits at the top of the vagina covering the cervix. This stops sperm from reaching the ovum in the fallopian tubes. The contraceptive sponge is another device that sits in the same place. These devices are covered with a chemical called spermicide, which kills sperm before they can reach the ovum. **Intrauterine devices** like the coil are placed inside the uterus by a doctor or nurse and prevent an embryo from implanting.

Another method of contraception for men is having a **vasectomy**. This minor surgical operation ties knots in or cuts the sperm ducts connecting the testes with the penis. So, no sperm are ejaculated into the vagina. Tying the ducts can be undone so this process is only temporary. A similar process called **tubal ligation** in women also exists. Again tying can be undone, but cutting the fallopian tubes is a permanent form of contraception. We call this sterilisation.

Condoms have an advantage over other forms of contraception. They stop the spread of many sexually transmitted diseases (STDs) because they prevent the exchange of any body fluids.

Intrauterine devices: Contraceptive devices, such as the coil, which are placed inside the uterus to prevent an embryo from implanting.

Vasectomy: A contraceptive medical procedure during which a man's sperm ducts are blocked or cut.

Tubal ligation: A contraceptive medical procedure during which a woman's fallopian tubes are blocked or cut.

The use of contraceptive hormones

The contraceptive pill contains the hormones oestrogen and progesterone. Women take the pill at the same time each day. This may be for 21 days of the cycle, followed by a week without any pills. Other types have pills for all 28 days, but the pills taken in the final week are just sugar. This method is often easier for women because they simply take a pill every day.

The pill (as it is often referred to) stops ovulation by reducing the production of follicle-stimulating hormone (FSH). It also thickens the mucus at the cervix and reduces the thickness of the uterus lining. Both of these processes reduce the chance of pregnancy. As well as preventing pregnancy, the contraceptive pill also helps women to keep their periods more regular and can make them lighter (less blood is released).

The same hormones can be delivered into a women's blood by contraceptive patches that stick to the skin and implants that are devices which sit underneath it. Patches last for 7 days and are used weekly for the first 3 weeks. An implant can function for several years before it needs to be replaced.

> **Exam tip**
>
> You should also be able to evaluate the different hormonal and non-hormonal methods of contraception.

The use of hormones to treat infertility

REVISED

As well as preventing pregnancy, hormones can be used to help it. Some women have naturally low levels of FSH and luteinising hormone (LH) hormones. Relatively simple injections of these hormones can help some women become pregnant.

In-vitro fertilisation (IVF) is often called having a test tube baby (although no test tubes are actually used). A woman is given injections of FSH and LH. Because of this, several ova mature. These are removed in a small operation and then mixed with the father's sperm. Fertilisation therefore happens outside of the body (*in vitro*). Shortly after fertilisation, the fertilised ovum divides into an embryo by mitosis. A second minor operation places fertilised ova back into the uterus of the mother. Not all ova will embed into the lining of the uterus, so often IVF treatments involve putting back multiple fertilised ova into the uterus. The chances of having twins during IVF is reasonably high.

Fertility treatments bring huge joy to people who might not otherwise be able to have children. However, these treatments can be very emotionally and physically stressful. The success rates are not often high. The possibility of multiple births can be a risk to both the babies and mother.

> **Exam tip**
>
> You should be able to explain the use of hormones to treat infertility.

Negative feedback control

REVISED

The regulation of blood glucose, water content and temperature are all examples of homeostasis. Negative feedback control is a crucial part of this process. This control occurs when your body detects a change, and then makes another change to return itself to normal.

Another example of negative feedback control involves the production of the hormone thyroxine in your **thyroid gland**. Your thyroid gland determines how quickly your body uses energy, makes proteins and how sensitive it is to other hormones. So it controls your metabolic rate. Your pituitary gland produces **thyroid-stimulating hormone (TSH)**, which in turn stimulates the release of thyroxine. This is an example of negative feedback control.

> **Thyroid gland:** A gland in your neck that produces thyroxine to regulate how quickly your body uses energy, makes proteins and how sensitive it is to other hormones.
>
> **Thyroid-stimulating hormone (TSH):** A hormone produced by your pituitary gland that regulates your thyroid gland.

(H) Just above your kidneys are your **adrenal glands**, which produce **adrenaline**. Unlike other hormonal responses, your body responds very quickly to adrenaline. It is produced when your body perceives a threat. It is often called the 'fight or flight' response. Adrenaline increases your heart rate and provides your muscles with more oxygen and glucose for respiration. This process allows your cells to respire more, releasing more energy for 'fight or flight' (run away).

> **Adrenal glands**: Glands found in your brain that produce adrenaline.
>
> **Adrenaline**: A hormone produced by your adrenal glands that causes an increase in heart rate ready for a fight or flight response.

Now test yourself

TESTED ☐

5 In which type of structures are hormones made?
6 Name your 'master gland' and its location.
7 What happens when blood glucose is too high?
8 How are the treatments for type 1 and type 2 diabetes different?
9 What four hormones control the menstrual cycle?
10 Which hormones are given to women before *in vitro* fertilisation?

Answers online

> **Exam tip**
>
> You should be able to explain the roles of thyroxine and adrenaline in the body.

Summary

- Homeostasis is the maintenance of a constant internal environment. These conditions are optimum for enzyme action and all cell functions. This includes controlling blood glucose concentration, body temperature (in warm-blooded animals) and water levels. Homeostasis can involve responses of the nervous system or hormones.
- Receptors are cells which detect changes. Coordination centres (such as the spinal cord, brain and pancreas) process information from receptors. Effectors are muscles or glands that bring about responses.
- A nerve is a bundle of neurone cells.
- The nervous system enables humans to respond to their surroundings and coordinate behaviour. Information from receptor cells passes along sensory neurones as electrical impulses to the central nervous system (CNS). The CNS is the brain and spinal cord. Within the CNS relay neurones transmit electrical impulses. Motor neurones carry electrical signals to effectors.
- Synapses are gaps between neurones which allow one neurone to pass a signal to several others. At a synapse, the electrical signal is converted into a chemical one in the form of neurotransmitters. These diffuse across the synapse and restart the electrical signal in the next neurone.
- In reflex reactions, signals are not immediately transferred to the conscious part of the brain. This saves time and potential damage to the body.

- The endocrine system is composed of glands that release hormones into the blood. They are carried to target organs where they have an effect. These signals are slower than responses of the nervous system.
- Body temperature is monitored by the thermoregulatory centre in the brain. If this is too high, blood vessels dilate (vasodilation) and sweat is produced from sweat glands. If it is too low, blood vessels constrict (vasoconstriction) and skeletal muscles contract to make you shiver.
- The pituitary gland in the brain is the master gland. Its secretes hormones which affect other glands. Other glands include the pancreas, thyroid, adrenal gland, ovaries and testes.
- Blood glucose is monitored and controlled by the pancreas. When too high, the pancreas produces the hormone insulin which converts glucose to glycogen in the liver and muscle cells.
- **(H)** When too low, the pancreas produces the hormone glucagon which converts glycogen in the liver and muscle cells to glucose. This forms a negative feedback cycle.
- Type 1 diabetes occurs when the pancreas fails to produce sufficient insulin. It is normally treated with insulin injections. Type 2 diabetes occurs when the body cells no longer respond to insulin that is produced. It is normally treated with a carbohydrate controlled diet and exercise. Obesity is a risk factor for type 2 diabetes.

→

- The menstrual cycle is an approximately 28-day reproductive cycle in women that runs from puberty to the menopause. Menstruation (having a period) occurs at the beginning of the cycle and last for several days. Following this, the lining of the uterus regrows in preparation for the embedding of a fertilised ovum if a woman is pregnant. Ovulation is the release of a mature ovum from an ovary which occurs around day 14. If a fertilised ovum does not embed, progesterone levels fall towards the end of the cycle and menstruation occurs.
- Follicle-stimulating hormone (FSH) causes an ovum to mature before ovulation. Luteinising hormone (LH) stimulates ovulation.
- The four hormones oestrogen, progesterone, FSH and LH interact to control the menstrual cycle.
- Contraception prevents pregnancy. Barrier contraception like condoms and diaphragms physically prevents sperm reaching the ovum. Condoms also stop the spread of STDs. Intrauterine devices prevent the implantation of the fertilised ovum. Spermicides are chemicals which kill sperm. Some people abstain from sex around ovulation. There are surgical operations for men and women which stop pregnancy.
- Hormonal contraceptives also prevent pregnancy. The contraceptive pill contains oestrogen and progesterone which inhibits FSH production which stops ova maturing. Injections, implants and patches have the same effect.
- FSH and LH can be given to women who are infertile to help them have a baby. *In vitro* fertilisation (IVF) involves the fertilisation of an ovum out of the body and its replacement into the uterus.
- Thyroxine from the thyroid gland stimulates metabolism. This is controlled by negative feedback. Adrenaline is a hormone produced by the adrenal glands. It boosts heart rate and the delivery of oxygen to the brain and muscles to prepare for a 'fight or flight' response.

Exam practice

1 Which of these is NOT a use of plant hormones? [1]
   A Selective insecticides
   B Producing seedless fruit
   C Rooting powder
   D Fruit ripening
2 Where is follicle-stimulating hormone (FSH) produced? [1]
   A Ovaries   B Pituitary gland   C Uterus   D Pancreas
3 Define homeostasis. [1]
4 Describe how hormones can be used for contraception. [3]
5 Describe what the results show for the experiments shown in the three separate diagrams. [3]

6 Describe the method used to investigate the speed of reaction times. [4]
7 Describe the structure and function of the nerves in the nervous system. [6]

Answers and quick quiz 5 online

ONLINE

6 Inheritance, variation and evolution

Reproduction

Sexual and asexual reproduction

Asexual reproduction

Asexual reproduction involves one parent organism that produces genetically identical offspring. These are called clones. There is no joining of gametes (sex cells), so there is no mix of DNA. Only mitosis is involved here. Many plants like strawberries and spider plants reproduce asexually when they produce tiny plantlets on **runners**. This is called vegetative reproduction. All bacteria reproduce asexually when they divide by **binary fission**. This process is like mitosis.

Sexual reproduction

Sexual reproduction involves two parents that produce genetically different offspring. The parent organisms produce gametes which then fuse during fertilisation. Gametes are ova and sperm in animals and ova and pollen in flowering plants. So, there is a mixing of genetic information which leads to variety in offspring.

Formation of gametes during **meiosis** requires energy. Some animals and plants produce gametes in incredibly high numbers. Many fish release millions of gametes at one time. The process of finding a mate (**courtship**) also often requires energy. Organisms that reproduce asexually do not need to use this energy.

Both sexual and asexual reproduction

Some organisms can reproduce both sexually and asexually depending upon the circumstances:

- The malarial **parasite** reproduces asexually in a human host but sexually in the mosquito.
- Many fungi can reproduce asexually by producing mitotic spores and sexually by producing meiotic spores.
- Many plants produce seeds sexually but also can reproduce asexually during vegetative reproduction.

Meiosis

Gametes are sex cells which are produced in meiosis. They are haploid and so must have half the number of chromosomes of a normal diploid body cell. A diploid human body cell like a nerve or muscle cell has 46 chromosomes, or 23 pairs. So, haploid human sperm and ova must have 23 chromosomes. During fertilisation, the sperm and ovum fuse to make a genetically different, diploid fertilised ovum. The two sets of 23 chromosomes have fused to form a new diploid cell with 23 pairs of chromosomes. This new cell divides by mitosis. Its cells can then differentiate as it grows into an adult organism.

Asexual reproduction: Reproduction involving one parent with genetically identical offspring.

Runner: An offshoot of a plant on which plantlets are produced by asexual reproduction.

Binary fission: The asexual reproduction of bacteria.

Meiosis: Cell division which forms non-identical, haploid gametes (sex cells)

Courtship: Behaviours to attract a mate.

Parasite: An organism that damages its host but depends on it to survive.

Exam tip

You should be able to explain the advantages and disadvantages of asexual and sexual reproduction.

Exam tip

You should be able to state that meiosis leads to non-identical cells being formed whilst mitosis leads to identical cells being formed.

The process of meiosis

The steps in the process of meiosis are shown in Figure 6.1.

1 Chromosomes make copies of themselves and nucleus disappears.

2 Chromosome pairs line up, and swap pieces of information (DNA crossover).

3 Cell divides.

4 Chromosomes line up.

5 Original and copied chromosomes move to opposite ends of the cell. Cell divides for a second time.

6 Four new nuclei form.

Figure 6.1 The main stages in meiosis for a cell with just two pairs of chromosomes. Once the matching chromosomes have been paired up, they may swap pieces of information between them. This is called DNA crossover. The cell divides twice and ends up as four daughter cells (gametes), each with half the original number of chromosomes.

During meiosis:

- Copies of the organism's DNA are made (a human cell would now have 92 chromosomes).
- The cell divides on two separate occasions (firstly back to two diploid cells with 46 chromosomes and then to four haploid cells with 23 chromosomes).
- DNA is exchanged to ensure all gametes are genetically different from each other.

Table 6.1 The key differences between mitosis and meiosis.

	Mitosis	Meiosis
Number of cells at beginning	One	One
Type of cell at beginning	Diploid body cell (23 pairs of chromosomes in humans)	Diploid body cell (23 pairs of chromosomes in humans)
Number of cells at end (daughter cells)	Two	Four
Type of cell at end	Diploid body cell (23 pairs of chromosomes in humans)	Haploid gamete (23 chromosomes in humans)
Number of divisions	One	Two
Identical or non-identical cells produced	Identical	Non-identical
Used for	Growth and repair	Producing gametes
Where it occurs	Everywhere except the sex organs	Sex organs (ovaries and testes in mammals)

Answers and quick quizzes at **www.hoddereducation.co.uk/myrevisionnotesdownloads**

DNA and the genome

Chromosomes and genes

Your genome is one copy of all your genetic information (your DNA). An identical copy of your genome exists in all your diploid cells because they were produced by mitosis. Your haploid gametes (sperm or ova) were produced by meiosis and so only have half of your genome.

Your genome consists of about 2 m of DNA. This fits inside the nucleus of most of your cells. (Your red blood cells do not have any DNA in them to maximise the oxygen they can carry.) To fit this length of DNA into your microscopic cells, it is arranged neatly into shapes called chromosomes. Your genome is made from 46 chromosomes which come in 23 pairs because half were present in each of the sperm and ova that made you. They paired up during fertilisation.

Chromosomes have regions that contain the DNA code to make proteins. These are called genes. Because you inherited one chromosome from each parent, you have two copies of almost all genes. These are called alleles. We think there are about 24 000 genes that carry the instructions to make our proteins. The rest of the DNA is non-coding. It does not make proteins and we call this 'junk DNA'.

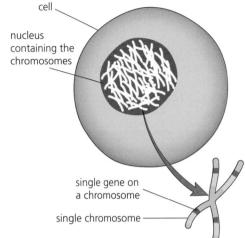

cell

nucleus containing the chromosomes

single gene on a chromosome

single chromosome

Figure 6.2 Make sure that you can identify the nucleus, chromosomes and the gene locations on the chromosomes.

The Human Genome Project

In 2003 a nine-year project finished which involved 20 universities in countries like the USA, UK, Japan, France, Germany and China. Collectively they identified every one of the around three billion bases that made up the genome of male and female volunteers. The results of this incredible effort are stored on the internet and are available to all.

Since its completion, scientists have identified the areas of the human genome that contain genes and those areas that are non-coding. From this project we have developed tests to show the likelihood of genetic disorders like **cystic fibrosis** occurring. It is highly likely that this will have great significance for medicine in the future. It is important that we understand the human genome so we can:
- search for genes linked to different types of disease
- understand and treat inherited disorders
- trace human **migration** patterns from the past.

Genetic inheritance

Sexual reproduction involves two parents who produce genetically different offspring. The offspring possess genes from both parents. Some characteristics are controlled by a single gene. Examples include eye colour and blood group in humans. You inherit a gene for each of these characteristics from each parent. So, you have two copies of each gene. We call these alleles.

Eye colour

Each of the sperm and the ovum that made you contained a gene for eye colour. These alleles determine your eye colour. If both genes gave the

> **Exam tip**
>
> You should be able to explain all of the key terms in this section.

same colour, say brown, it makes sense that you would have brown eyes. But what if they were different? At GCSE you only need to learn about brown and blue eyes. We use letters to represent the colours. We call these **genotypes**. **Phenotypes** are a description of a genotype using words.

Because you inherited a gene from each of your parents there are two letters in a genotype. Brown eyes are **dominant** over blue eyes. We use a capital letter to show this. Those genes that can be dominated are called **recessive**. So B is a brown eyed gene and b is a blue eyed gene. Inheriting a B from both parents (BB) is called **homozygous dominant**. Inheriting a b from both parents (bb) is called **homozygous recessive**. Inheriting one of each (Bb) is called **heterozygous**.

Table 6.2 The three possible allele combinations (genotypes) for eye colour.

Genotype	Phenotype	Terminology
BB	Brown eyes	Homozygous dominant
bb	Blue eyes	Homozygous recessive
Bb	Brown eyes	Heterozygous

We complete genetic crosses in **Punnett squares** to see the likelihood of inheriting certain characteristics. In the example in Figure 6.3 the mother's genotype is BB. This means her phenotype is brown eyes. Because she is homozygous dominant, all her ova will have B. The father's genotype is bb. Because he is homozygous recessive all his sperm will have b. So, each of the four possible combinations for their children will be Bb. That is, all their children will be heterozygous and have brown eyes.

The outcomes of these crosses can be given as percentages or ratios.

The possibilities of other characteristics such as whether you have ear lobes or can roll your tongue can be determined using Punnett squares. These are shown in Table 6.3.

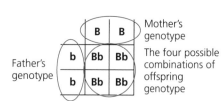

Figure 6.3 The four possible genotypic offspring of the parents whose genotypes were BB and bb.

Table 6.3 The three possible genotypes and phenotypes for ear lobes and tongue rolling.

Terminology	Ear genotype	Ear phenotype	Tongue genotype	Tongue phenotype
Homozygous dominant	EE	Free lobes	TT	Can roll
Homozygous recessive	ee	Attached lobes	tt	Can't roll
Heterozygous	Ee	Free lobes	Tt	Can roll

Most characteristics are because of multiple genes interacting, rather than a single gene as described previously.

Genotype: The genetic make-up of an organism represented by letters.

Phenotype: The physical characteristics of an organism as described by words.

Dominant: Will show a characteristic if inherited from one or both parents.

Recessive: Will show a characteristic only if inherited from both parents.

Homozygous dominant: A genotype with two dominant alleles.

Homozygous recessive: A genotype with two recessive alleles.

Heterozygous: A genotype with one dominant and one recessive allele.

Punnett square: A grid that makes determining the chance of inheriting a characteristic easier to understand.

Revision activity

Draw out these tables with only the headings along the top and the first column on the left. Try to fill in the rest of the tables from memory to help you to revise.

Exam tip

Be careful when filling in Punnett squares that your capital and lower case letters look different. A 'C' and a 'c' (for example) can look very similar.

Family trees

A family tree can show the inheritance of characteristics over multiple generations. Every generation has its own horizontal line, with the oldest at the top.

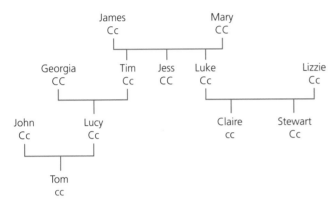

Figure 6.4 The gene for cystic fibrosis is shown by the letter 'c' in this family tree. CC is normal. Cc is a carrier who does not have the disorder but who could give it to their children. cc is a person with cystic fibrosis.

Inherited disorders

REVISED

Communicable diseases are caused by pathogens. Inherited conditions, like cystic fibrosis, are called disorders and the people who inherit them from their parents are called sufferers.

Cystic fibrosis

About one in every 10 000 people in the UK has cystic fibrosis. Sufferers inherit a recessive allele from both parents. So, all sufferers are homozygous recessive. If they had inherited one dominant gene from either parent, they would be heterozygous and not have cystic fibrosis. In this case, they would be able to pass it to their children and so we call them carriers for the disorder.

Cystic fibrosis is a disorder of cell membranes. Sufferers have excess mucus produced in their lungs, digestive and reproductive systems. This often becomes infected. Frequent physiotherapy sessions remove much of the mucus. Sadly, there is no cure for cystic fibrosis and sufferers have a much-reduced life expectancy.

Polydactyly

Polydactyly is a genetic disorder that results in an extra toe or finger. This is very rare condition that results from the inheritance of a dominant allele from one or both parents.

Sex determination

We call the 23rd pair of chromosomes the sex chromosomes because this is what they determine. Again, here we use letters. All ova are X. Approximately half of sperm are also X and the rest are Y. An X ovum and an X sperm develop into a female. An X ovum and a Y sperm develop into a male.

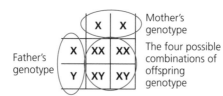

Mother's genotype

The four possible combinations of offspring genotype

Father's genotype

Figure 6.5 The relative proportions of offspring of this cross explain why approximately 50 percent of the human population is of each sex.

Exam tip

You should be able to carry out a genetic cross to show sex inheritance. You should be able to use direct proportions and simple ratios.

Now test yourself

1 What are the products of asexual reproduction?
2 How do strawberries and spider plants reproduce asexually?
3 Define the term selective breeding.
4 What are the products of meiosis in animals?
5 What are two key differences of the products of meiosis and mitosis?
6 Why is it important in medicine that the human genome is understood?
H▶7 What percentage of children will be homozygous recessive for an inherited disorder if both parents are carriers?

Answers online

Variation

Variation

Variation is the sum of all the differences between two organisms of the same or different species. So, it is all the differences between two cats (the same species) and also a cat and a dog.

Causes of variation

Variation can be caused by:
1 Environmental factors such as scars and tattoos
2 Genetic factors like blood group and eye colour
3 Both environmental and genetic factors together for your weight and height.

When genetic and environmental factors work together, a person's genome interacts with the environment to influence their development.

Types of variation

The results from investigations into variation can be grouped into two types: continuous and discontinuous. **Continuous data** comes in a range and values can be halfway between them. Height is an example of this. You can be 140, 141 or 140.5 cm tall. Continuous data are presented in a line graph with a line of best fit. **Discontinuous data** comes in discrete

Variation: The differences that exist within a species or between different species.

Continuous (data): Data that come in a range and not in groups.

Exam tip

You should be able to describe how genetics (a person's genome) interact with the environment to influence the development of organisms.

Discontinuous (data): Data that come in groups and not a range.

groups. Your blood group can be A, B, AB or O. It cannot be halfway between them. Discontinuous data are presented in a bar chart.

Normally distributed variation

Line graphs of continuous data often show a characteristic 'bell-shaped graph'. This is shown in Figure 6.6. We call this a **normal distribution**. It means most values are towards the middle, and there are an ever-smaller number towards the outsides.

Mutations

A change to our DNA is called a mutation. These mutations occur naturally when our cells divide by mitosis. They occur more commonly in cells exposed to carcinogenic chemicals or ionising radiation. Very rarely this exposure can change the characteristics of organisms in a population. If these changes are advantageous, this can lead to a rapid evolutionary change.

> **Exam tip**
>
> You should be able to recall that most mutations have little or no effect on an organism's phenotype.

> **Normal distribution**: Data that are more common around a mean and form a bell-shaped graph.

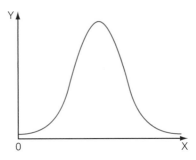

Figure 6.6 Bell-shaped graphs show normally distributed data. The most common values are in the middle and the least common values are at both ends.

Selective breeding

REVISED

All dogs are one species. That means that, regardless of their breed, they can all interbreed to produce fertile offspring. All dogs are also descended from wolves. They have not evolved from wolves but been selectively bred by humans. Before they understood the mechanisms of inheritance or the process of **evolution**, our ancestors knew that if they bred a large dog with a large bitch, they were likely to get large puppies. This selective breeding has occurred repeatedly over thousands of years and has resulted in large breeds like the Great Dane. Similarly, breeding a protective dog and bitch over many generations has led to the development of breeds like the German Shepherd. This is **selective breeding**. It is also called **artificial selection** to remind us that it is not natural selection, which leads to evolution.

Selective breeding has given us Jersey cows, which have been selected to produce creamy milk and Friesian cows, which produce a larger volume of less creamy milk. We have also selectively bred:
- crops that are resistant to disease
- animals that produce more meat
- domestic dogs that have a gentle nature
- large and unusual flowers.

Selective breeding can lead to a reduction in the variation of a population. This is called **inbreeding** and results in genetic weakening of species. Occasionally, and by mistake, the selection of a key characteristic like size can also magnify a less desirable one. Many pedigree dogs suffer from hip misalignment because of this.

> **Exam tip**
>
> You should be able to explain the impact of selective breeding of food plants and domesticated animals.

> **Evolution**: The theory first proposed by Charles Darwin that the different species found today formed as a result of the accumulation of small advantages that were passed through generations.
>
> **Selective breeding**: Breeding animals or plants with desirable characteristics.
>
> **Artificial selection**: As in selective breeding.
>
> **Inbreeding**: Artificial selection of a small number of parents, which reduces variation.

> **Revision activity**
>
> Are you clear about the differences between artificial selection in selective breeding and natural selection in evolution?

Genetic engineering

Genetic engineering is a modern and technical process by which the genome of an organism is altered by adding a gene from another organism. This allows us to directly transfer desired characteristics into species. Organisms altered in this way are called genetically modified (GM) or **transgenic**.

This is much quicker than selective breeding. Genetic engineering is sometimes called **genetic modification**. It is an ethical issue, which means some people disagree with it for religious or moral reasons. There are many regulations around genetic engineering to tightly control it. It is illegal to genetically engineer humans, but modern medical research is exploring the possibility of using genetic engineering to overcome some inherited disorders.

Glow-in-the dark rabbits

Genetic engineering has inserted the gene that makes jellyfish glow-in-the-dark into rabbits. Enzymes were used to cut out the specific gene from the genome of the jellyfish. The genome of a rabbit embryo was then cut open using the same enzyme. Another enzyme is used to seal the glow-in-the-dark gene into the embryo. The embryo was then implanted back into the uterus of a rabbit to grow normally.

It is much harder to insert the gene into every cell of an adult organism. So, we choose to genetically engineer an embryo, which will then divide naturally into an adult organism. All cells will contain the inserted gene.

Genetically engineered crops

We have genetically engineered crops to:
- be resistant to disease
- be resistant to being eaten by insects or herbivores
- produce larger yields (e.g. bigger, better fruits).

Golden rice has been genetically engineered to contain carotene. This reduces the chance of vitamin A deficiency, which causes blindness in children. Cotton has been genetically modified to be resistant to an insect pest called the weevil. Soya has been genetically modified to be herbicide resistant.

Many people think that genetic engineering of crops has the potential to feed starving people, particularly in those countries that experience drought or **famine**. Other people think that we should not interfere with God's creatures. Others worry about the spread of genes from genetically engineering crops into wild species. We are not yet sure of the effects of GM crops on wild flowers and insects. Some people are also concerned that we have not fully explored the effects of GM crops on human health.

Genetically engineered animals

Sheep have been genetically engineered to produce proteins, such as blood clotting factors that we use in medicine, in their milk. This process is shown in Figure 6.7.

Genetic engineering: A scientific technique in which a gene is moved from one species to another.

Transgenic: Describes a genetically engineered organism.

Genetic modification: As genetic engineering.

Famine: An extreme shortage of food, often leading to many deaths.

Exam tip

You should be able to describe genetic engineering as a process that involves modifying the genome of an organism by introducing a gene from another organism to give a desired characteristic.

Exam tip

You should be able to describe the main steps in genetic engineering.

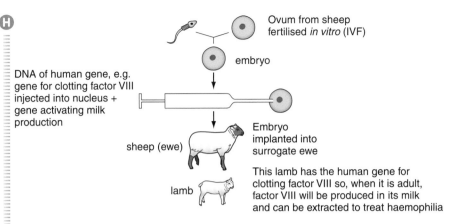

DNA of human gene, e.g. gene for clotting factor VIII injected into nucleus + gene activating milk production

Ovum from sheep fertilised *in vitro* (IVF)

embryo

sheep (ewe)

Embryo implanted into surrogate ewe

This lamb has the human gene for clotting factor VIII so, when it is adult, factor VIII will be produced in its milk and can be extracted to treat haemophilia

lamb

Figure 6.7 Genetically engineered sheep can produce human proteins in their milk.

Genetically engineered micro-organisms

We have genetically modified bacterial cells to contain the human gene for insulin. As these bacteria grow, they produce human insulin which we collect. Diabetics inject this.

> **Exam tip**
>
> You should be able to explain the potential benefits and risks of genetic engineering in agriculture and medicine.

Now test yourself

TESTED

8 Describe the causes of variation.
9 State two examples of environmental variation.
10 Define the term continuous data.
11 What type of distribution is observed in a bell-shaped graph?
12 What process have farmers used to produce Friesian and Jersey cows?
13 What is genetic engineering?
14 Describe why we have genetically modified crops.
H▶15 How, specifically, has genetic modification been used to treat haemophilia?

Answers online

The development of understanding of genetics and evolution

The theory of evolution

REVISED

Evolution explains how the millions of different species alive today and those that have already become extinct developed from one **common ancestor**, which was alive more than 3 billion years ago. It explains how over many generations tiny changes in individuals give them an advantage and allow them to develop to better suit their surroundings. Eventually these differences add up to make a new species. This occurs when the changes mean that the two populations can no longer interbreed to have fertile offspring.

> **Common ancestor**: An organism from which others have evolved.

Evidence for evolution

REVISED

Since Darwin published his theory of evolution, evidence for this process has developed to the point that almost all scientists agree with it. Most criticism of it now comes from religious groups who believe in **creationism**. This is the belief that God created the Universe and all life in it, within 7 days.

Fossils

Fossils are the remains of dead organisms preserved for millions of years in rock. Many fossils are preserved when their tissues are replaced by minerals as they decay. Other fossils are the tracks or traces of organisms. These can include dinosaur footprints, burrows and eggs. In special places like **peat** bogs, conditions are such that the rate of decay is very slow. This preserves fossils extremely well.

The **fossil record** is the information provided by all the fossils that have ever been discovered. This record can show us how significant the changes have been to species over time. There are however, gaps in the fossil record because not all fossils have been found, many have been destroyed by magma and not all parts of soft-bodied organisms become fossils.

Creationism: A belief that God created all the organisms on Earth and not evolution.

Peat: Partially decayed vegetation.

Fossil record: All of the fossils that have been discovered so far.

Antibiotic-resistant bacteria

Sir Alexander Fleming (1881–1955) discovered the first antibiotic called penicillin in 1928. Since then we have developed several other antibiotics. This process is very slow. It seems that bacteria are evolving to become immune to our antibiotics faster than we can develop new ones. A common strain of bacteria that has developed resistance is **MRSA (methicillin-resistant *Staphylococcus aureus*)**. MRSA is a communicable pathogen that kills several hundred people per year in the UK. To reduce the speed at which antibiotic resistance develops we should reduce the use of antibiotics generally and always finish the full course of medicine we are given.

Because bacteria reproduce much more quickly than many other animals, we can see the evolution of antibiotic resistance in our own lifetimes. This change provides strong evidence for evolution.

Methicillin-resistant *Staphylococcus aureus* (MRSA): A bacterium that has evolved to be resistant to antibiotics.

Exam tip

You should be able to describe the evidence for evolution including fossils and antibiotic-resistant bacteria.

Extinction

REVISED

Evolution explains that organisms with characteristics most suited to the environment are more likely to survive and breed successfully. The reverse is also true. Those without these adaptations are less likely and eventually may become extinct. This occurs when there are no remaining individuals of a species still alive.

There have been several points in the Earth's history when many species have become extinct in a short period. These are called **mass extinction** events. The asteroid that killed the dinosaurs is an example of this. The latest mass extinction event is being caused by us. The rate at which extinctions are currently occurring is increasing together with an increase in the human population. We are cutting down rainforests, **overfishing** our oceans, creating bigger cities and farming more intensively. It seems that every year more species are being added to the endangered list.

Mass extinction: A large number of extinctions occurring at the same time (humans are the latest cause of a mass extinction).

Overfishing: Fishing on a scale so large that the population of species is threatened.

Exam tip

You should be able to describe factors which may contribute to the extinction of a species.

Now test yourself

TESTED

16 Define the term common ancestor.
17 Describe Darwin's theory of evolution.
18 What are fossils and how do they provide evidence for evolution?
19 What is the significance of MRSA bacteria?
20 What has been the ultimate fate of organisms that were not well adapted to their environment?

Answers online

Classification of living organisms

Classification

REVISED

Classification is the process by which things are placed into groups based upon their characteristics. Scientific classification of the different species of life is a very important process. Without classification, we would not be able to identify organisms. We would not know how many there were in a species. So, we would not know which ones needed conversation to protect them from becoming extinct.

Carl Linnaeus and binomial classification

Classification began with Carl Linnaeus (1707–1778). He put things into their groups based upon their structure and characteristics. He developed the **binomial** system of classification in which all living species are given a two-part name. The first is their **genus** and second their species. The binomial name for humans is *Homo sapiens*. *Homo* is our genus and *sapiens* is our species. Thousands of years ago we shared our planet with other very closely related species called *Homo neanderthal* and *Homo erectus* that are now extinct. The fact that the three species are in the same genus means they are very closely related.

A species is a group of organisms that can interbreed to produce fertile offspring. All dogs are one species so they can all interbreed. Occasionally some very closely related species can interbreed but they produce infertile **hybrids** as offspring. Horses and donkeys produce infertile animals called mules. Lions and tigers produce infertile animals called ligers.

The five-kingdom system

Linnaeus's system of binomial classification puts all life into ever bigger groups. There are five large groups called **kingdoms**. These are animals, plants, fungi, bacteria and protists. Each of these is split into smaller groups called phyla, then classes, orders and families, before genera (the plural of genus) and species.

The three-domain system

As equipment like electron microscopes and techniques involving genetic mapping of genomes has developed, so has our ability to classify organisms. Carl Woese (1928–2012) pioneered these

> **Binomial**: Having two names; a genus and a species (e.g. *Homo sapiens*).
>
> **Genus**: The second smallest group of classifying organisms.
>
> **Hybrid**: An infertile organism produced when two different species interbreed.
>
> **Kingdom**: The largest group of classifying organisms, e.g. the animal kingdom.

Typical mistake

It is important to remember that the first part of any binomial name is the genus and the second is the species.

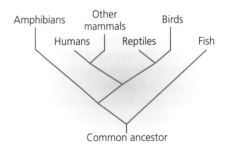

Figure 6.8 An evolutionary tree showing the relationship of some vertebrates.

developments and proposed the three-domain system. In this, organisms are classified into three main groups:

1 Archaea (primitive bacteria usually living in **extreme environments**)
2 Bacteria (**prokaryotes**)
3 Eukaryotes (protists, plants, fungi and animals).

Extreme environment: A location in which it is challenging for most organisms to live.

Prokaryotes: Prokaryotic organisms (bacteria).

Evolutionary trees

After they have been classified, the relationship between different species can be shown in diagrams called evolutionary trees.

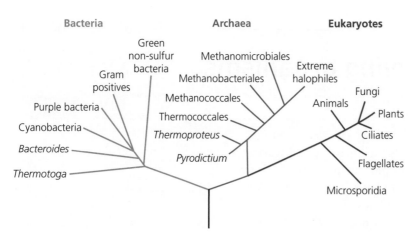

Figure 6.9 The evolutionary tree devised by Woese and co-workers

Exam tip

You should be able to describe the impact of developments in biology on classification systems.

Now test yourself

TESTED ☐

21 Define the term binomial classification.
22 State the binomial name for humans.
23 Which scientist developed the five-kingdom model of classification?
24 State the five kingdoms.
25 Describe the hierarchy of classification from kingdom to species.
26 What are hybrids?
27 What led to the development of the three-domain system from the five kingdoms?
28 What are the three domains?
29 In what environments are Archaea found?
30 Which scientist developed the three-domain model?

Answers online

Summary

- Like mitosis, meiosis is a type of cell division. Unlike mitosis, it produces four non-identical cells with half the DNA of the original cell. These are gametes (sex cells), which are sperm and ova in animals and pollen and ova in flowering plants.
- During meiosis, the genetic information is firstly copied. The cell then divides twice to form four non-identical daughter cells.
- Sexual reproduction involves the fusion of a male and a female gamete to restore the normal number of chromosomes. So, there is

a mixing of genetic information which leads to variety in offspring. This is called fertilisation in animals and pollination in plants.
- Asexual reproduction requires one parent (not two) and so there is no fusion of gametes. Offspring are genetically identical clones of the parent. Only mitosis, not meiosis, is involved.
- DNA is a polymer made from two strands forming a double helix. In eukaryotes, this is arranged into chromosomes. A gene is a section of DNA that codes for a sequence of amino acids to make a specific protein.

- Your genome is all your genetic material. The entire human genome has been sequenced in the Human Genome Project. This collaborative work was completed by 20 universities in different countries. The results are available for all on the internet. This is an important first step in the search for genes linked to different types of disease, understanding and treatment of inherited disorders, and use in tracing human migration patterns from the past.
- Some characteristics are controlled by one gene such as fur colour in mice and red–green colour blindness in humans. Other examples include inheritance of eye colour, tongue rolling and the presence of ear lobes. However, most characteristics are controlled by the interaction of more than one gene.
- Alleles are pairs of genes, one inherited from each parent. The alleles present are an organism's genotype. Their expression is an organism's phenotype. A dominant allele is always expressed even if only one allele is present. A recessive allele is only expressed if two copies are present. The presence of two dominant or recessive alleles is called homozygous and the presence of one dominant and one recessive allele is called heterozygous.
- The possible outcomes of genetic crosses can be shown in Punnett squares. Inheritance of characteristics and disorders in families can be shown in family trees.
- Some medical disorders are inherited. Polydactyl is an anomaly that causes having extra fingers or toes and is caused by a dominant allele. Cystic fibrosis sufferers produce excess mucus in their respiratory, digestive and reproductive systems. This condition is caused by a recessive allele.
- Sex is determined by the combination of the 23 pairs of chromosomes. All ova are X. Half of sperm are X and half are Y. XX is female and XY is male.
- Variation is the sum of all the differences between two organisms of the same or different species. The causes of variation are genetic (e.g. eye colour and blood group), environmental (e.g. scars and tattoos) and combinations of both (e.g. height and weight).
- Data resulting from surveys into variation come in a range (continuous) or groups (discontinuous). Many biological features are normally distributed and give characteristic bell-shaped graphs.
- There is usually extensive variation within a species. This arises from mutations. Most of these have no effect on the phenotype, some influence and very few determine phenotype. Advantageous mutations can lead to relatively quick changes in species.

- Evolution is a change in the inherited characteristics of a population over time through the process of natural selection which may result in the formation of a new species (speciation). This occurs when the phenotypes of two populations mean they can no longer interbreed.
- Selective breeding occurs when humans select individual plants and animals to breed for their particular genetic characteristics. This is repeated over many generations. Examples include disease resistance in food crops, animals which produce more meat or milk, domestic dogs with a gentle nature and large or unusual flowers. Selective breeding can lead to inbreeding, which can magnify the incidence of disease or inherited defects.
- Genetic engineering is the process of modifying the genome of an organism by the introduction of a gene from another organism to give a desired characteristic. Plant crops have been genetically engineered to be resistant to disease and increase yield. Bacterial cells have been genetically engineered to produce human insulin to treat diabetes. Genetic engineering is an ethical issue.
- (H) In genetic engineering, enzymes are used to isolate and cut out the required gene. This gene is then inserted into a bacterial plasmid or virus. This vector is then used to insert the gene into the cells of animals, plants or micro-organisms so they develop desired characteristics.
- Charles Darwin developed his theory of evolution by natural selection during an around-the-world expedition. He delayed publication of his theory because of his concern about the reaction of the Church.
- Fossils are the 'remains' of organisms from millions of years ago. The fossil record provides evidence for evolution. There are gaps in the fossil record because many early forms of life were soft-bodied, not all fossils have been found and many have been destroyed by geological processes. The development of antibiotic-resistant bacteria like MRSA is also evidence for evolution.
- Extinction occurs when there are no remaining individuals of a species still alive.
- To reduce the speed at which bacteria are becoming antibiotic resistant, doctors should not prescribe antibiotics unnecessarily, patients should complete their full course and the use of antibiotics in agriculture should be restricted.
- Classification is the process of grouping organisms. It is an essential step before conservation of endangered organisms can occur.
- Classification began with Carl Linnaeus in the 18th century. He classified species into a kingdom, phylum, class, order, family and

→

genus. He developed the binomial system of naming organisms. The first part of the name is the genus and the second part is the species.

● As a result of advances in microscope and DNA sequencing technology, Carl Woese proposed his 'three-domain' system. Organisms are either *archaea* (primitive bacteria), bacteria or eukaryotes.

● Evolutionary trees show the relationship between the evolution of organisms.

Exam practice

1 What cause of variation are scars and tattoos examples of? [1]
   A Continuous
   B Genetic
   C Discontinuous
   D Environmental

2 What does the second part of a binomial name indicate? [1]
   A Kingdom
   B Class
   C Genus
   D Species

3 Complete this Punnett Square for eye colour. [2]

		Female alleles	
		B	b
Male alleles	B		
	b		

4 Describe the differences in the cells produced in mitosis and meiosis. [4]

5 What does this image tell you about the evolutionary relationships between the six classes of vertebrate? [4]

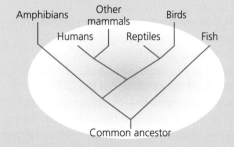

6 Draw a graph of the results from a survey into blood groups shown in the table below. [6]

Blood group	A	B	AB	O
Number of people	21	5	2	24

7 A man heterozygous for eye colour marries a homozygous recessive woman. Complete a Punnett square to show their possible offspring. Analyse your results as percentages and proportions. [6]

	Female alleles		
Male alleles			

8 Doctors are now prescribing fewer antibiotics to reduce the evolution of antibiotic-resistant bacteria. Describe the process of evolution of antibiotic bacteria. [6]

Answers and quick quiz 6 online

ONLINE

7 Ecology

Adaptations, interdependence and competition

Communities

A **population** is the total number of all the organisms of the same species in a geographical area. This can either be the whole planet or a much smaller part of it. A **community** is a group of two or more populations of different species in the same geographical area.

An **ecosystem** is the interaction between a community of living organisms and the non-living parts of their environment. To survive and reproduce, all organisms need resources from their surroundings and the other organisms that live with them.

In all communities, populations of organisms compete for resources. If this competition is within one population of organisms it is called intraspecific **competition**. If it is between two populations, we call it interspecific competition. Plants compete for light, space, water and nutrients from the soil. Animals compete for food, mates and territory.

> **Population**: The total number of all the organisms of the same species or the same group of species that live in a particular geographical area.
>
> **Community**: A group of two or more populations of different species that live at the same time in the same geographical area.
>
> **Ecosystem**: A community of living organisms in their environment.
>
> **Competition**: The contest between organisms for resources such as food and shelter.

Interdependence

All organisms within a community depend upon each other. This can be for food, shelter, pollination and seed dispersal. They have evolved to do this. We call this **interdependence**. A stable community is one in which high levels of interdependence are found. Here we see balance between predators and prey. The numbers rise and fall, as shown in Figure 7.1. But interdependence means they never get to the point at which one organism kills or totally outcompetes another. Removal of one species from an ecosystem can affect all other organisms within it.

> **Interdependence**: All the organisms in a community depend upon each other and because of this changes to them or their environment can cause unforeseen damage.

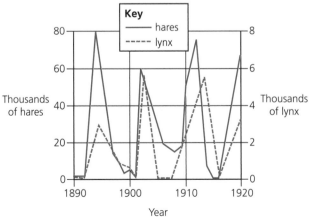

Exam tip

You should be able to describe the importance of interdependence and competition in a community.

Figure 7.1 Predator–prey cycling in the Canadian lynx and snowshoe hare populations.

Abiotic factors

REVISED

Abiotic factors are non-living. They can be chemical or physical, but not biological. Important abiotic factors for plants and animals are: light intensity for photosynthesis, temperature, moisture levels, soil pH and mineral content, wind intensity and direction and carbon dioxide for plants and oxygen levels for aquatic animals.

> **Abiotic factors**: The non-living parts of the environment.

Exam tip

You should be able to explain how a change in an abiotic factor could affect a community.

Biotic factors

REVISED

Biotic factors are living. Important biotic factors include the availability of food, numbers of new predators, introduction of a pathogen that causes a communicable disease and competition between species. Biotic factors often result from newly introduced species. If these factors damage the local ecosystem by outcompeting existing species, they are called **invasive species**. The cane toad in Australia and grey squirrel in the UK are invasive species.

> **Biotic factors**: The living parts of the environment.
>
> **Invasive species**: An organism that is not native and causes negative effects.

Exam tip

You should be able to explain how a change in a biotic factor could affect a community.

Adaptations

REVISED

All organisms have adaptations that enable them to survive in the conditions in which they normally live. They allow individuals to outcompete others and provide them with an evolutionary advantage. Without this and the resulting competition, there would be no evolution. Adaptations can be structural, behavioural or physiological.

Structural adaptations are physical features that allow competition. They include good eyesight, sharp teeth or claws and shells or exoskeletons for protection. **Behavioural adaptations** are specific behaviours that give an organism an advantage. Animals like crows, primates, elephants and alligators use tools. This is an obvious behavioural advantage. **Physiological adaptations** are processes that help organisms. The formation of poisons by some snakes, spiders and plants is an example.

> **Structural adaptation**: An advantage to an organism as a result of the way it is formed, like the streamlining seen in fish.
>
> **Behavioural adaptation**: An advantage to an organism as a result of behaviour, such as a courtship display.
>
> **Physiological adaptation**: An advantage to an organism as a result of a process, such as the production of poisonous venom.

Exam tip

You should be able to suggest how organisms are adapted to the conditions in which they live.

Extreme environments

Extreme environments have conditions in which most life finds it hard to survive. They have:
- extremes of pH or temperature
- low levels of oxygen or water
- high salt concentration.

Typical mistake

It is important that you can explain why an organism is adapted not just state its adaptation.

Answers and quick quizzes at **www.hoddereducation.co.uk/myrevisionnotesdownloads**

Deserts are extreme environments because they are very hot during the day and cold at night. They are also extremely dry. The polar regions are also extreme environments because of their cold temperature, low light level in winter and the lack of fresh, liquid water.

Deep-sea **hydrothermal vents** are gaps in the seabed where magma provides heat. Surrounding them are communities of life which exist nowhere else on Earth. These are the only food chains that do not start with a photosynthesising plant or alga. Bacteria feed on the chemicals from the vents and support all other species of life. Organisms that live in extreme environments are called **extremophiles**. Near the vents, the temperatures are very high but further away they very quickly fall. Hydrothermal vents are always dark and under high pressure.

Hydrothermal vents: Volcanic vents at the bottoms of seas and oceans, where unique species of life have evolved based upon bacteria feeding on chemicals and not photosynthesis.

Extremophile: An organism that lives in an extreme environment.

Now test yourself

TESTED ☐

1 Define the term population.
2 Define the term community.
3 What do animals compete for?
4 What is interdependence?
5 State two examples of abiotic factors that could affect the distribution of plants in the environment.
6 What biotic factors could affect the distribution of animals in the environment.
7 State an example of a physiological adaptation.
8 What conditions exist in extreme environments?
9 What term do we give to organisms that live in extreme environments?
10 What is unusual about food chains that surround hydrothermal vents?

Answers online

Organisation of an ecosystem

A population is the total number of all the organisms of the same species in a geographical area. A community is a group of two or more populations of different species in the same geographical area. An ecosystem is the interaction between a community of living organisms and the non-living parts of their environment. A habitat is the natural environment in which an organism lives.

Sampling

REVISED ☐

Sampling is the process of looking at a small part of an ecosystem and drawing conclusions about the whole. This saves important time and money.

Sampling: The process of recording a smaller amount of information to make wider conclusions.

Quadrat: A square frame used in biological sampling.

Quadrats

Quadrats are squares of wire that are often 0.25 m². They are often used during sampling to record the number of organisms in a specific area (within the quadrat). Quadrats are used in three ways:

- The number of a single species within them is counted.
- The number of different species are counted (a measure of biodiversity).
- The percentage cover of one species such as grass is recorded.

There are two main ways in which you use quadrats depending upon what it is you are trying to investigate. If you want to know the numbers of a species in an area, or to compare two or more areas you would place your quadrats randomly. If you want to investigate the change in a habitat you would place your quadrats systematically along a line called a **transect**.

In all cases of sampling, it is important you record more than three concordant results. It is likely you will place at least 20 quadrats and calculate mean values from your results.

Random sampling using quadrats

You cannot stand in the middle of the habitat you wish to sample and throw your quadrat. It is important that the placement is totally random to avoid bias in your results. The method you use is described below in Required practical 7 'Sampling'.

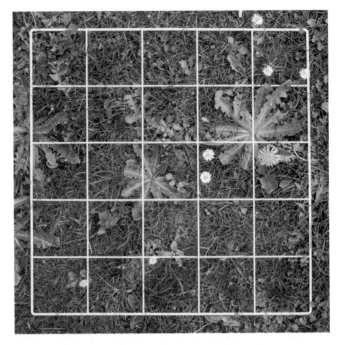

Figure 7.2 A quadrat on a lawn with weeds to count.

> **Transect**: A line along which systematic sampling occurs.
>
> **Systematic sampling**: The regular distribution (not random) of a survey to answer a specific question, usually about a trend.

Required practical 7

Sampling

Aim: To measure the population size of a common species in a habitat. Use sampling techniques to investigate the effect of a factor on the distribution of this species.

Equipment: Quadrat, ruler

Method:
1 Choose a starting location in one corner of your ecosystem.
2 Generate random numbers using a table or calculator.
3 Use two random numbers as coordinates to tell you where to place your first quadrat.
4 Record the total number of individuals of the species you are investigating in this quadrat.
5 Return to your original starting position and repeat using different random numbers for coordinates.

Results:
Calculate a mean value for your results.

If you have counted the number of a species within your quadrat, you can estimate how many are in your habitat. To do this:
● Measure the area of your habitat.
● Multiply the mean number of organisms per quadrat by the difference in size between your quadrat and habitat.

Systematic sampling using quadrats

Systematic sampling looks for changes in the distribution of organisms as a result of changes within a habitat. Because of this, other abiotic factors like light intensity or moisture levels are often recorded alongside biotic ones.

If you wanted to see if the number and species of seaweed changed as you walked down a seashore you would use systematic sampling. (This is different from the total number of seaweed on the shore, for which you could use random sampling.) Draw an imaginary line called a transect down your habitat. Place your quadrat at equal and regular distances down the transect. Record the number and species of seaweed in each quadrat. At the same time, record abiotic factors that might help explain the changes you observe. (In the case of the seashore, the position of seaweed is determined largely by the number of hours any species can be out of water. Those at the top of the shore have evolved to be out of the water for longer.)

Exam tip

You should be able to explain the principles of sampling.

Producers, consumers and decomposers

REVISED

All organisms need energy to complete the seven life processes: movement, reproduction, sensitivity, nutrition, excretion, respiration and growth.

Producers

Producers are photosynthetic plants and algae found at the lowest **trophic level** of almost all food chains. During photosynthesis, these species turn carbon dioxide and water to glucose and oxygen using sunlight. This glucose then supports all life at higher trophic levels. Unusually, bacteria in deep-sea hydrothermal vents are also producers. They feed on the chemicals released from the volcanic vents, which are in turn consumed by all other species in these very specific food chains. Producers are called **autotrophs** because they 'feed' themselves.

Consumers

Consumers are organisms that obtain their energy by eating others. All animals are consumers. Any that eat photosynthetic plants or algae are called primary consumers. They are also called herbivores. Consumers that eat primary consumers are called secondary consumers. Then comes tertiary and quaternary consumers as we move up the trophic levels. At the top of all food chains is the **apex predator**. Consumers that kill and eat other animals are predators and those that are eaten are prey. Consumers obtain their food from other organisms and are called **heterotrophs**.

Interestingly, at each trophic level only 10% of the energy from the previous one is passed along. So, when cows eat grass, or lions eat zebras they only obtain 10% of the total energy. The rest is used by the organisms (grass or zebras) to complete the seven life processes.

Producer: Any organism that photosynthesises (a plant or alga).

Trophic level: A stage in a feeding relationship representing an organism in a food chain or a group of organisms in a food web.

Autotroph: An organism that makes its food from simple organic compounds in its surroundings, often using energy from light.

Consumer: Any organism in a feeding relationship that eats other organisms for food.

Apex predator: The final organism in a feeding relationship.

Heterotroph: An organism that obtains its food from other organisms, e.g. humans.

Exam tip

You should be able to recall that photosynthetic organisms are the producers of most biomass for life on Earth.

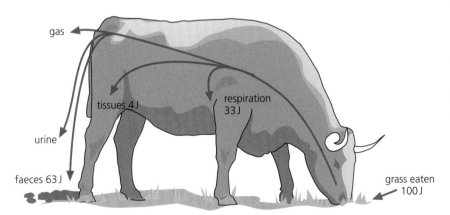

gas

tissues 4 J

respiration 33 J

urine

faeces 63 J

grass eaten 100 J

Figure 7.3 The energy use of a cow. Compare the food energy eaten with the amount built into body tissue. Look at the large amount of energy left in faeces. What organisms can use this?

Materials recycling

All materials in the living world are recycled to provide the building blocks for future organisms. That is, some of your atoms were incorporated into previous living organisms. They will also be incorporated into other organisms after you die.

The carbon cycle

The carbon cycle shows the various carbon compounds that it can form and how it is converted between them. It returns carbon from organisms to the atmosphere as carbon dioxide to be used by plants in photosynthesis.

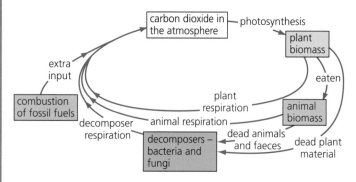

Figure 7.4 The carbon cycle.

The three key processes of the carbon cycle are shown in Table 7.1.

Table 7.1 The key processes and conversions of the carbon cycle.

Process	Carbon starts as	Carbon ends as
Photosynthesis	Carbon dioxide	Glucose
Respiration	Glucose	Carbon dioxide
Combustion (burning)	Fuel	Carbon dioxide

Combustion: Burning.

> **Revision activity**
>
> Draw out this table with only the headings along the top and the first column on the left. Try to fill in the rest of the table from memory to help you to revise.

The water cycle

The water cycle shows all the pathways water takes as it cycles through our atmosphere, rivers, lakes, seas and oceans. It provides fresh water for plants and animals on land before draining into the seas.

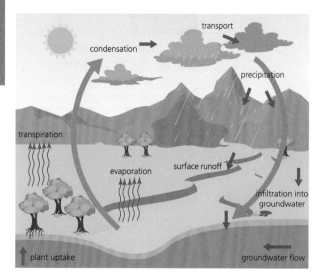

Figure 7.5 The water cycle.

Precipitation is the scientific name for rain, snow, hail and sleet. **Runoff** occurs when water moves across the surface of land. This is often in streams and rivers. **Infiltration** is the movement of water downwards from the surface to become groundwater. Evaporation is the conversion of liquid water to gas. Water evaporates from all surface water. Transpiration is the movement of water though plants from roots to its leaves. Water vapour evaporates from tiny pores in leaves that are called stomata.

Precipitation: Rain, snow, hail and sleet.

Runoff: The movement of water across the surface of land.

Infiltration: The movement of water into the ground to become groundwater.

Now test yourself

TESTED

11 Define the term sampling.
12 State the three ways in which quadrats are used.
13 When would a transect be used instead of random sampling?
14 What term describes an organism that makes its food from simple organic compounds in its surroundings, often using energy from light.
15 What proportion of energy is transferred between tropic levels?
16 State the three main processes of the carbon cycle.
17 What is precipitation?

Answers online

Exam tip

You should be able to evaluate the impact of environmental changes on the distribution of species in an ecosystem.

Biodiversity and the effect of human interaction on ecosystems

Biodiversity

REVISED

Biodiversity is a measure of the variety of all the different species of organisms on Earth, or within a particular ecosystem. Areas of low biodiversity include deserts and the polar regions. Areas of high biodiversity include tropical rainforests and ancient oak woodlands.

A relatively high biodiversity for any ecosystem ensures its stability by reducing the dependence of one species on another for food, shelter and the maintenance of the physical environment. Crucially, the future of the human species on Earth probably depends upon us maintaining a good level of biodiversity. Many of our activities, including **deforestation**, pollution and burning of fossil fuels, reduce biodiversity. Are we doing enough?

Deforestation: Cutting down of trees (often on a large scale).

Waste management

REVISED

The human population has increased past seven billion. One hundred years ago, there was less than two billion people. This rapid growth in our population and an increase in our standard of living has meant that we are using evermore resources. We are also producing evermore waste. Unless our waste is properly disposed of, evermore pollution will be produced.

Water pollution

All life needs water. We used water for drinking, growing our food, washing, transportation and to find food. Pollution of water hurts us and the animals and plants that live in or near this water that we also depend upon. It is estimated that over one billion people do not currently have access to clean water.

Water pollution comes from pathogens like *Salmonella* bacteria, *Norovirus* and parasitic worms. We often find these in water polluted by sewage. Fertilisers can also wash from farmer's fields in heavy rains and pollute nearby streams and rivers. This can lead to **eutrophication**. Some factories still illegally release toxic chemicals into rivers and oceans. Some of these, like the **pesticide** DDT and metal mercury, cannot be easily excreted. So they concentrate at higher trophic levels. We call this **bioaccumulation**. Oil spills have drastically polluted some coastal areas killing hundreds of thousands of sea birds, otters and seals.

Air pollution

Air pollution is often caused by waste gases from vehicles or factories. Without suitable regulations, excessive particulates produced from these sources can cause **smogs** that are seen covering large cities in some parts of the world. Sulfurous oxides are produced by the burning of fossil fuels. They can react with water vapour to form **acid rain**. They can destroy entire forests of trees and damage stone buildings and statues. Carbon monoxide is another polluting gas. This poisonous gas is produced during **incomplete combustion**. It is odourless and replaces oxygen in your red blood cells. It slowly suffocates you.

Land pollution

In recent years, our society has become more consumable. We seem to be mending fewer things and just buying more to replace them. This change has resulted in a significant increase in the rubbish we produce. Despite the efforts of local councils, a significant amount of this rubbish is still not recycled and goes into landfill. As our waste slowly rots away it can produce a toxic liquid called leachate and release large volumes of greenhouse gases. To reduce landfill, it is essential that we reduce our consumption, reuse as much as we can before buying new and **recycle**. (These are called the three R's.)

Land use

REVISED

The way we use our land is called land use. This can be for farming crops (**arable**) or rearing animals. It can be forests of trees to produce timber. It can be urban for towns and cities. We quarry or mine some areas and dump waste in others. It can also be preserved in national parks or other conservation areas. During the Stone Age, much of the UK was covered in forests and very little was farmland. Think about the UK now. The clear majority of it is now farmland or urban towns and cities. Our land use has changed dramatically in this time.

Peat bogs

Peat bogs have wet, acidic soils with low levels of nutrients. Very few trees grow in them. They have relatively low **biodiversity** but contain species that are not found in other places. Because of these conditions, very little decomposition occurs in them and peat accumulates. It is partially decayed vegetation. Peat can be the first step in the slow process of forming fossil fuels. In some parts of the world, peat is dug up and used as a fuel. This releases lots of carbon dioxide into the atmosphere that was stored as a carbon '**sink**' in the peat. The destruction of peat bogs reduces biodiversity.

Eutrophication: Death of all life in an aquatic ecosystem as a result of overuse of fertilisers often on nearby farmland.

Pesticides: Chemicals used to kill pests.

Bioaccumulation: The increase in concentration of toxins at higher trophic levels in a food chain.

Smog: Fog or haze because of smoke or other polluting gases.

Acid rain: Precipitation that is acidic because of air pollution.

Incomplete combustion: The burning of fuel without sufficient oxygen, which produces poisonous carbon monoxide.

Recycle: Changing a waste product into a new raw material to make another.

Arable: Farming of crops for food.

Biodiversity: A measure of the different species present in a community.

Sink: A long-term store of a substance, often carbon.

Deforestation

Deforestation is the cutting down of trees so that an area can be used for other purposes, often to make farmland for cattle or rice fields, or grow biofuels. Deforestation began around 12 000 years ago as we turned from **hunter-gatherers** to farmers. However, in recent years the rate at which deforestation has occurred has increased massively. Now huge areas of rainforest are cut down to grow crops like palm oil. It therefore drastically reduces biodiversity. We have cut down over half of the rainforest that existed 75 years ago.

Deforestation stops plants photosynthesising and so they do not remove carbon dioxide from the atmosphere. The wood from deforestation is often burned which releases more carbon dioxide into the atmosphere. Deforestation therefore increases the greenhouse effect and global warming.

> **Hunter-gatherer:** A member of a nomadic tribe who live without farming but by hunting, fishing and collecting wild food.

Global warming

Global warming is the gradual increase in the Earth's average temperature. This has changed naturally over time. We have had ice ages and other times when tropical conditions were present in the UK. However, almost all scientists agree that the current rate of change is faster than the planet has ever seen before. They also almost all agree that this is occurring because of the greenhouse effect.

The effects of global warming include the melting of glaciers and polar icecaps, raising ocean levels threatening low lying cities like London and New York, freak weather and species migration. How long will it be before global warming threatens **food security** in the UK?

> **Exam tip**
>
> You should be able to describe some of the biological consequences of global warming.

> **Food security:** How safe the supply of our food is.

Greenhouse effect

Greenhouse gases include carbon dioxide and methane. As we release larger volumes of greenhouse gases, they are trapping more heat in our atmosphere. At low levels, the greenhouse effect is needed to keep our planet warm enough to support life. However, analysis of drilled ice cores from the polar regions has shown our human activity in the last few hundred years has massively increased the carbon dioxide in our atmosphere. It is essential that our governments, companies and voluntary organisations, and we as individuals, all act to stop this increase before it is too late.

Maintaining biodiversity

Biodiversity is a measure of the numbers of different species on our planet or in an ecosystem. **Conservation** is one way in which biodiversity is maintained. This protects certain areas with high ecological importance like peat bogs, ancient forests and marine regions like coral reefs. Other ways of maintaining biodiversity include:

- zoo **breeding programmes** to increase the number of endangered species
- reintroduction of hedgerows in areas that have been **intensively farmed**
- reducing deforestation and carbon dioxide emissions by some governments
- reducing, reusing and recycling rather than dumping in landfill.

> **Conservation:** Protecting an ecosystem or species of organism from reduced numbers and often extinction.
>
> **Breeding programme:** Activity of zoos to breed captive animals together to increase their gene pool.
>
> **Intensive farming:** Industrial agriculture to maximise yield, often involving the use of machines, chemical fertilisers and pesticides.

Now test yourself

18 What is biodiversity?
19 Why do some pesticides bioaccumulate?
20 What process produces poisonous carbon monoxide?
21 What three 'R's' help live a more sustainable lifestyle?
22 What conditions are found in peat bogs?
23 What are carbon 'sinks'?
24 Describe the effects of global warming.
25 How is biodiversity maintained?
26 Define the term food security.
27 What is intensive farming?

Answers online

Exam tip

You should be able to describe both positive and negative human interactions in an ecosystem and their impacts on biodiversity.

Summary

- A population is the total number of all the organisms of the same species in a geographical area. A community is a group of two or more populations of different species in the same geographical area. An ecosystem is the interaction between a community of living organisms and the non-living parts of their environment.
- In all communities, populations of organisms compete for resources. Plants compete for light, space, water and nutrients from the soil. Animals compete for food, mates and territory.
- Interdependence means all organisms within a community depend upon each other.
- Abiotic factors are non-living and include light intensity, temperature, moisture levels, soil pH and mineral content, wind intensity and direction, carbon dioxide levels for plants and oxygen levels for aquatic animals.
- Biotic factors are living and include availability of food, new predators arriving, new pathogens and one species outcompeting another.
- Both abiotic and biotic factors affect the distribution of organisms in an ecosystem.
- Adaptations are features which allow organisms to survive in the conditions in which they normally live. These can be structural, behavioural or physiological.
- Extreme environments include areas with high temperatures, pressures or salt concentration. The polar regions and deep-sea hydrothermal vents are examples. Organisms that live in these conditions are called extremophiles.
- Photosynthesising plants and algae are the producers of nearly all biomass (living tissue) on Earth. Food chains and webs show the feeding relationships of organisms within a community. A producer (usually a plant or alga) is found at the first trophic level. Producers are eaten by primary consumers (herbivores) which are in turn eaten by secondary and tertiary consumers (carnivores).
- The numbers and distribution of organisms in an ecosystem can be investigated by using quadrats and transects (imaginary lines along which quadrats are placed).
- Many different materials, including water and carbon, cycle through abiotic and biotic parts of an ecosystem. The carbon and water cycles are important for all living organisms. The carbon cycle returns carbon from organisms to the atmosphere as carbon dioxide to be used by plants for photosynthesis. Micro-organisms cycle materials through an ecosystem by returning carbon to the atmosphere and mineral ions to the soil. The water cycle provides fresh water for plants and animals on land before draining into the sea.
- Biodiversity is the variety of all the different species of organisms on Earth or within an ecosystem. High levels of biodiversity ensure stable ecosystems. The future of humanity on Earth probably depends upon us maintaining high biodiversity.
- A rapid recent increase in the human population and an increase in the standard of living mean we are producing more waste. If not treated appropriately, waste becomes pollution. This pollution can occur in water by sewage, fertilisers or toxic chemicals. It can occur in air from smoke and acidic gases. It can also occur on land from landfill and from toxic chemicals. Pollution reduces biodiversity.
- Building, quarrying, farming and dumping waste in landfill reduce the land available for other species. The destruction of peat bogs further reduces biodiversity.
- Deforestation has occurred to provide land for cattle and rice fields and to grow crops for biofuels.
- Global warming is caused by the greenhouse effect. Carbon dioxide and methane are greenhouse gases.
- Biodiversity is maintained by breeding programmes, protection and regeneration of rare habitats, reintroducing hedges, the reduction of deforestation and carbon dioxide emissions, and recycling.

Exam practice

1 Which of these is not an abiotic factor?　　　　[1]
   - A Light intensity
   - B Temperature
   - C Water availability
   - D Disease
2 What are transects?　　　　[1]
   - A Square frames of wire used for sampling
   - B Imaginary lines along which sampling occurs
   - C Small containers used to suck up insects
   - D Large nets used to sweep though plants to collect insects
3 Define the term community.　　　　[1]
4 Describe the conditions near a deep-sea hydrothermal vent.　　　　[2]
5 Many young mammals can stand and run within hours of being born. Explain why this is an advantage.　　　　[2]
6 Describe and explain the predator–prey cycling shown for hares and lynx.　　　　[6]

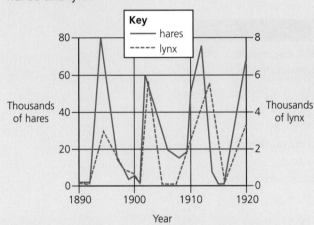

7 Describe the processes in the water cycle.　　　　[6]
8 Describe the method you would use see if there are more species of plant on the school field where it is cut or left uncut.　　　　[6]

Answers and quick quiz 7 online

ONLINE

8 Atomic structure and the periodic table

Atomic structure

The structure of atoms

- Atoms are the smallest part of an element that can exist.
- Atoms are very small. Typical atoms have a radius of about 0.1 nm (1×10^{-10} m).
- Atoms are made from smaller particles called protons, neutrons and electrons.

$1 \text{ nm} = 1 \times 10^{-9} \text{ m}$

Table 8.1

	proton	neutron	electron
where it is	in the nucleus	in the nucleus	outside the nucleus in energy levels (shells)
relative charge	+1	0	–1
relative mass	1	1	very small

- The protons and neutrons are contained in a tiny central nucleus.
- The nucleus is about 10 000 times smaller than the atom.
- The electrons move around the nucleus in energy levels (also called shells).

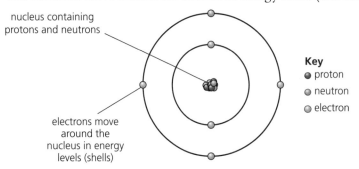

nucleus containing protons and neutrons

electrons move around the nucleus in energy levels (shells)

Key
- proton
- neutron
- electron

Figure 8.1

Atomic number and mass number

- The number of protons, neutrons and electrons in an atom is found from the mass number and atomic number.
- Atoms are neutral and so the number of protons equals the number of electrons.

Atomic number number of protons

Mass number number of protons + number of neutrons

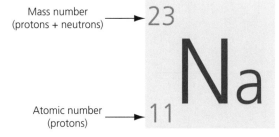

Mass number (protons + neutrons)

23

Na

11

Atomic number (protons)

Figure 8.2

Table 8.2

atom	atomic number	mass number	protons = atomic number	neutrons = mass number – atomic number	electrons = atomic number (only for atoms but not ions)
$^{23}_{11}Na$	11	23	11	12	11
$^{40}_{18}Ar$	18	40	18	22	18
$^{35}_{17}Cl$	17	35	17	18	17
$^{37}_{17}Cl$	17	37	17	20	17

these two atoms are isotopes

Isotopes and relative atomic mass

REVISED

- **Isotopes** are atoms of the same element that have a different mass number (e.g. $^{35}_{17}Cl$ and $^{37}_{17}Cl$).
- This means that isotopes are atoms with the same number of protons but a different number of neutrons.
- Isotopes of the same element have the same chemical properties because they have the same electron structure.
- The relative atomic mass (A_r) of an element is the average mass of all the isotopes of an element.

$$\text{relative atomic mass } (A_r) = \frac{\text{total mass of all the atoms of an element}}{\text{total number of atoms of an element}}$$

> **Isotopes** atoms with the same number of protons but a different number of neutrons

Example

Find the relative atomic mass of chlorine which consists of 75% $^{35}_{17}Cl$ and 25% $^{37}_{17}Cl$ atoms.

$$\text{relative atomic mass } (A_r) = \frac{(75 \times 35) + (25 \times 37)}{75 + 25} = 35.5$$

> **Exam tip**
>
> The relative atomic mass must have a value between the mass of the heaviest and the lightest isotopes. For example, for chlorine the value must be between 35 and 37. If your answer is outside this range, you have made a mistake.

Electron structure

REVISED

- The electrons orbit the nucleus in energy levels (shells).
- Electrons are in the lowest possible energy level (the energy levels closest to the nucleus).
- The 1st energy level holds a maximum of 2 electrons and the 2nd energy level holds a maximum of 8 electrons. The next 8 electrons go into the 3rd energy level, with the next 2 electrons going into the 4th energy level.

Table 8.3

atom	aluminium	potassium
number of electrons	13 electrons	19 electrons
diagram of electron structure		
written electron structure	2,8,3	2,8,8,1

8 Atomic structure and the periodic table

AQA GCSE (9–1) Combined Science Trilogy 87

Ions

- **Ions** are particles with an electric charge because they contain a different number of protons and electrons.
- Positive ions have more protons than electrons.
- Negative ions have more electrons than protons.
- Common ions have the same electron structure as the Group 0 elements (the noble gases). These are very stable electron structures. (The only common ion that is an is exception is H^+ which has no electrons at all.)

> **Ion** charged particle with a different number of protons and electrons

> **Typical mistake**
>
> Many candidates struggle to work out the number of electrons in an ion. Remember that protons are positive and electrons are negative.
>
> An ion with a 1+ charge has more positive protons than negative electrons, for example, the $^7_3Li^+$ ion has 3 protons (charge = 3+) and therefore must have 2 electrons (charge = 2−).

Table 8.4

ion	atomic number	mass number	protons	neutrons	electrons	electron structure
$^7_3Li^+$	3	7	3	4	2	2
$^{27}_{13}Al^{3+}$	13	27	13	14	10	2,8
$^{35}_{17}Cl^-$	17	35	17	18	18	2,8,8
$^{16}_8O^{2-}$	8	16	8	8	10	2,8

Development of ideas about atomic structure

- As scientists make new discoveries, scientific models have to be changed or completely replaced.
- For many years, scientists thought that atoms were the smallest possible particle that could not be divided.
- This theory was replaced by the plum-pudding model of the atom when electrons were discovered. In this model, the atom is a ball of positive charge with negative electrons embedded in it.
- The plum-pudding model was replaced a few years later when the results of an alpha particle scattering experiment showed that it could not be correct.
- In the experiment, alpha particles were fired at a very thin piece of gold foil. If the plum-pudding model was correct, the alpha particles would have passed straight through the gold foil. However, a few alpha particles were deflected or bounced back.
- Rutherford worked out from these results that atoms have a tiny, positive nucleus surrounded by electrons. Most of the mass of the atom is concentrated in this nucleus. This is known as the nuclear model.

plum-pudding model (1897)

nuclear model (1911, but revised since)

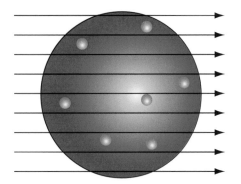
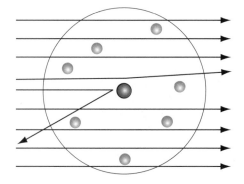

The alpha particles would all be expected to travel straight through the gold foil according to the plum-pudding model

A tiny fraction of alpha particles were deflected or bounced back. Rutherford worked out that there must be a tiny, positive nucleus to explain this

Figure 8.3 The alpha scattering experiment

Answers and quick quizzes at **www.hoddereducation.co.uk/myrevisionnotesdownloads**

- Bohr revised the nuclear model when he showed that the electrons moved in energy levels.
- Later experiments showed that the positive nucleus is made of particles called protons.
- A few years later, the nuclear model was revised again when Chadwick discovered that neutrons are in the nucleus as well as the protons.

Now test yourself

TESTED

1 What is the atomic number and mass number of an atom with 26 protons, 30 neutrons and 26 electrons?
2 How many protons, neutrons and electrons are there in each of these atoms and ions?
(a) ${}^{9}_{4}Be$ (b) ${}^{39}_{19}K$ (c) ${}^{24}_{12}Mg^{2+}$ (d) ${}^{19}_{9}F^{-}$
3 What is the electron structure of each of the atoms and ions in question 2?
4 What is the charge on an ion containing 16 protons and 18 electrons?
5 What is special about the electron structure of ions?
6 Explain why ${}^{12}_{6}C$ and ${}^{13}_{6}C$ are isotopes.
7 Calculate the relative atomic mass of gallium which contains 60% ${}^{69}_{31}Ga$ atoms and 40% ${}^{71}_{31}Ga$ atoms. Give your answer to 3 significant figures.
8 The radius of a calcium atom is 0.18 nm. State this in standard form in metres.

Answers online

Reactions of elements

Elements in the periodic table

REVISED

- An element is a substance containing one type of atom. For example, all the atoms in the element carbon are carbon atoms.
- Elements cannot be broken down into simpler substances.
- There are just over 100 elements.
- Each element has its own symbol (the first letter of which is always a capital).
- The elements are listed in order of atomic number in the periodic table.

Typical mistake

It is very important to show the capital letters and small letters clearly when writing symbols. Some students can lose marks by mixing up capital or small letters or not writing them clearly.

Group 1	Group 2											Group 3	Group 4	Group 5	Group 6	Group 7	Group 0
							H 1										He 2
Li 3	Be 4											B 5	C 6	N 7	O 8	F 9	Ne 10
Na 11	Mg 12											Al 13	Si 14	P 15	S 16	Cl 17	Ar 18
K 19	Ca 20	Sc 21	Ti 22	V 23	Cr 24	Mn 25	Fe 26	Co 27	Ni 28	Cu 29	Zn 30	Ga 31	Ge 32	As 33	Se 34	Br 35	Kr 36
Rb 37	Sr 38	Y 39	Zr 40	Nb 41	Mo 42	Tc 43	Ru 44	Rh 45	Pd 46	Ag 47	Cd 48	In 49	Sn 50	Sb 51	Te 52	I 53	Xe 54
Cs 55	Ba 56	La* 57	Hf 72	Ta 73	W 74	Re 75	Os 76	Ir 77	Pt 78	Au 79	Hg 80	Ti 81	Pb 82	Bi 83	Po 84	At 85	Rn 86
Fr 87	Ra 88	Ac** 89	Rf 104	Db 105	Sg 106	Bh 107	Hs 108	Mt 109	Ds 110	Rg 111	Cn 112	Uut 113	Fl 114	Uup 115	Lv 116	Uus 117	Uuo 118
			*	Ce 58	Pr 59	Nd 60	Pm 61	Sm 62	Eu 63	Gd 64	Tb 65	Dy 66	Ho 67	Er 68	Tm 69	Yb 70	Lu 71
			**	Th 90	Pa 91	U 92	Np 93	Pu 94	Am 95	Cm 96	Bk 97	Cf 98	Es 99	Fm 100	Md 101	No 102	Lr 103

Metal
Non-metal
Difficult to classify

Figure 8.4 Elements are listed in order of atomic number in the periodic table

Element a substance that cannot be broken down into simpler substances

Compound a substance made from two or more different elements bonded together

● Over three quarters of the elements are metals. Most of the other elements are non-metals, but a few are hard to classify.

Table 8.5 Physical properties of metals and non-metals

	metals	non-metals
melting and boiling points	high (usually)	low (usually)
conductivity	thermal and electrical conductor	thermal and electrical insulator
density	high density (usually)	low density
appearance	shiny when polished	dull (usually)
malleability	can be hammered into shape	brittle as solids

Reactions between elements

REVISED

● Compounds are formed when elements react with each other.
● When elements react, electrons are transferred or shared so that atoms obtain the stable electron structure of the noble gases (Group 0 elements).

Table 8.6 Reactions between metals and non-metals

	metal + non-metal	metal + metal	non-metal + non-metal
Is there a reaction?	reacts	no reaction	reacts
What happens with electrons	electrons are transferred from metal to non-metal		electrons are shared
Type of compound formed	ionic		molecular (covalent)

Balancing equations

REVISED

● Word equations show the names of the **reactants** (the chemicals at the start) and the **products** (the chemicals made in the reaction).
● A balanced equation shows the formula of each substance and how many particles of each are involved in the reaction.

Reactants the chemicals at the start of a reaction

Products the chemicals made in a reaction

Table 8.7

	reactants	products	what the equation tells us
word equation	hydrogen + oxygen $\rightarrow$ water		hydrogen reacts with oxygen to form water
balanced equation	$2H_2 + O_2 \rightarrow 2H_2O$		2 molecules of H_2 react with 1 molecule of O_2 to form 2 molecules of H_2O

● In a balanced equation, the total number of atoms of each element in the reactants and products must be the same. This is because atoms cannot be created or destroyed.

- How to balance an equation:

  1 Write a word equation (you may be given this).   e.g methane + oxygen ⟶ carbon dioxide + water

  2 Re-write the equation with the formula of each substance in place of the name (you may be given the equation at this stage and be asked to balance it).

  3 Count the number of each atom in the reactants and products to see if it is balanced.

  4 If the equation is not balanced, then add in extra molecules. NEVER change the formula of a substance.

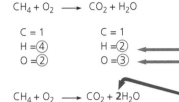

$CH_4 + O_2 \longrightarrow CO_2 + H_2O$

C = 1	C = 1
H = ④	H = ②
O = ②	O = ③

this is not balanced as the number of H and O atoms is different on each side

$CH_4 + O_2 \longrightarrow CO_2 + 2H_2O$

C = 1	C = 1
H = 4	H = 4
O = ②	O = ④

add another H_2O molecule on the right so that there are 4 H atoms on both sides – but the O atoms are still not balanced

add another O_2 molecule on the left so that there are 4 O atoms on both sides – the equation is now BALANCED

$CH_4 + 2O_2 \longrightarrow CO_2 + 2H_2O$

C = 1	C = 1
H = 4	H = 4
O = 4	O = 4

- Sometimes balanced equations include state symbols which are shown:
  (s) = solid
  (l) = liquid
  (g) = gas
  (aq) = aqueous (i.e. dissolved in water)

Now test yourself

TESTED ☐

9 What type of compound, if any, will be formed when the following elements react with each other?
  (a) sodium + bromine
  (b) oxygen + sulfur
  (c) magnesium + aluminium
  (d) chlorine + iron
10 Balance the following equations.
  (a) $Mg + O_2 \rightarrow MgO$
  (b) $Na + H_2O \rightarrow NaOH + H_2$
  (c) $Fe_2O_3 + C \rightarrow Fe + CO$
11 Write in words what the following equation tells us:
  $P_4(s) + 5O_2(g) \rightarrow P_4O_{10}(s)$

Answers online

Typical mistake

Students are sometimes tempted to change the formula of a substance to balance an equation. You should **never** do this because it changes what the substances actually are.

The periodic table

Electron structure and the periodic table

REVISED ☐

- All the elements in the same group in the periodic table have the same number of electrons in their outer shell. For example, all the elements in group 7 have 7 electrons in their outer shell.
- Elements in the same group have similar chemical properties because they have the same number of electrons in their outer shell.

Group 1	Group 2														Group 3	Group 4	Group 5	Group 6	Group 7	Group 0
							H 1													**He** 2
Li 2,1	**Be** 2,2														**B** 2,3	**C** 2,4	**N** 2,5	**O** 2,6	**F** 2,7	**Ne** 2,8
Na 2,8,1	**Mg** 2,8,2														**Al** 2,8,3	**Si** 2,8,4	**P** 2,8,5	**S** 2,8,6	**Cl** 2,8,7	**Ar** 2,8,8
K 2,8,8,1	**Ca** 2,8,8,2																			

Figure 8.5 The electron structure of the first 20 elements in the periodic table

Group 0 – the noble gases

- The noble gases include helium (He), neon (Ne) and argon (Ar).
- They are all non-metals.
- They are very unreactive because they have stable electron structures.
- They are all colourless gases at room temperatures.
- They have very low boiling points as their atoms are not bonded together and there are only very weak forces between the atoms.
- Going down the group, the boiling points increase as the atoms get heavier and the forces between the atoms increase.

Now test yourself

12 How many electrons are there in the outer shell of the following elements?
   (a) calcium   (b) nitrogen   (c) potassium   (d) neon   (e) sulfur
13 Explain why magnesium and calcium have similar chemical properties.
14 Explain why argon is very unreactive.

Answers online

Group 1 – the alkali metals

Group 1

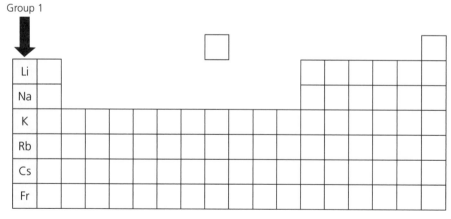

Figure 8.6

- The alkali metals include lithium (Li), sodium (Na) and potassium (K).
- Each one is a soft metal with a low density and a low melting point.
- They are very reactive as they all have one electron in their outer shell which is easily given away.
- When they react, they give away the outer shell electron forming 1+ ions (e.g. Li^+, Na^+, K^+) in an ionic compound.

Table 8.8

	reaction with oxygen	reaction with chlorine	reaction with water
lithium (Li)	burns with a crimson red flame to form a white powder $4Li + O_2 \rightarrow 2Li_2O$	burns with a crimson red flame to form a white powder $2Li + Cl_2 \rightarrow 2LiCl$	moves around on the surface of the water and fizzes $2Li + 2H_2O \rightarrow 2LiOH + H_2$
sodium (Na)	burns with a yellow-orange flame to form a white powder $4Na + O_2 \rightarrow 2Na_2O$	burns with a yellow-orange flame to form a white powder $2Na + Cl_2 \rightarrow 2NaCl$	melts, moves around on the surface of the water, fizzes $2Na + 2H_2O \rightarrow 2NaOH + H_2$
potassium (K)	burns with a lilac flame to form a white powder $4K + O_2 \rightarrow 2K_2O$	burns with a lilac flame to form a white powder $2K + Cl_2 \rightarrow 2KCl$	melts, moves around on the surface of the water, fizzes, catches fire with a lilac flame $2K + 2H_2O \rightarrow 2KOH + H_2$

- When alkali metals react with water, a solution of a metal hydroxide (e.g. sodium hydroxide) and hydrogen gas are formed. The metal hydroxides are alkalis.

Trend in reactivity

- The elements get **more** reactive going down the group. This is because:
  - the outer electron is further from the nucleus
  - the attraction between the outer electron and the nucleus is weaker
  - the outer electron is lost more easily.

lithium (2, 1)   sodium (2, 8, 1)

Figure 8.7 The further down group 1, the further the outer electron is from the nucleus

Now test yourself

TESTED

15 Write a word equation for the reaction of sodium with water.
16 Is sodium or potassium more reactive? Explain your answer.
17 What is the charge on the rubidium ions formed when rubidium reacts with oxygen?

Answers online

Group 7 – the halogens

REVISED

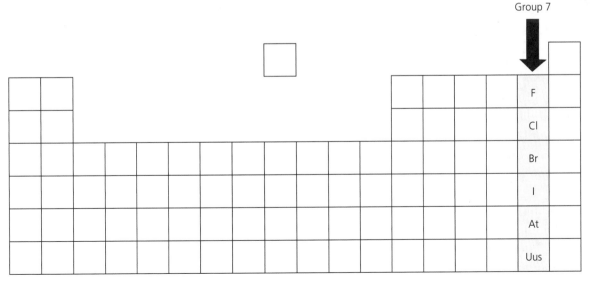

Figure 8.8

- The halogens include fluorine (F_2), chlorine (Cl_2), bromine (Br_2) and iodine (I_2).
- They are all non-metals.
- Each is made of diatomic molecules, i.e. molecules containing two atoms.
- They have low melting and boiling points because there are weak forces between the molecules.

Table 8.9

halogen	fluorine	chlorine	bromine	iodine
formula of molecules	F_2	Cl_2	Br_2	I_2
appearance	yellow gas	green gas	orange-brown liquid	grey solid

- Going down the group, the melting and boiling points increase because the forces between the molecules get stronger.

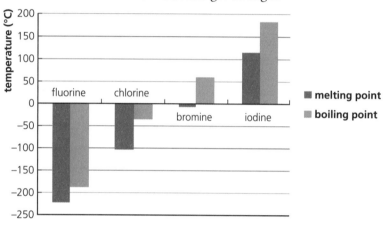

Figure 8.9

- The elements have 7 electrons in their outer shell.
- When they react, they gain an extra electron by either:
  ○ forming 1– ions when they react with metals
  ○ forming covalent bonds by sharing electrons when they react with non-metals.

Trend in reactivity

Table 8.10 Halogen displacement reactions

	chlorine water, Cl_2(aq) (very pale green solution)	bromine water, Br_2(aq) (yellow solution)	iodine water, I_2(aq) (brown solution)
sodium chloride solution, NaCl(aq) (colourless solution)		no reaction	no reaction
sodium bromide solution, NaBr(aq) (colourless solution)	**yellow solution forms** as chlorine displaces bromine (as chlorine is more reactive than bromine) $Cl_2 + 2NaBr \rightarrow 2NaCl + Br_2$ $Cl_2 + 2Br^- \rightarrow 2Cl^- + Br_2$		no reaction
sodium iodide solution, NaI(aq) (colourless solution)	**brown solution forms** as chlorine displaces iodine (as chlorine is more reactive than iodine) $Cl_2 + 2NaI \rightarrow 2NaCl + I_2$ (H)► $Cl_2 + 2I^- \rightarrow 2Cl^- + I_2$	**brown solution forms** as bromine displaces iodine (as bromine is more reactive than iodine) $Br_2 + 2NaI \rightarrow 2NaBr + I_2$ (H)► $Br_2 + 2I^- \rightarrow 2Br^- + I_2$	

- The elements get **less** reactive going down the group. This is because:
  - it is harder to gain an electron
  - the electron gained is further from the nucleus
  - the attraction between the electron gained and the nucleus is weaker.
- The trend in reactivity can be shown by halogen displacement reactions.
- A more reactive halogen will displace a less reactive halogen from its compounds.

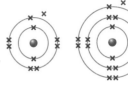

fluorine (2,7)   chlorine (2,8,7)

Moving down the group, the electron gained is further from the nucleus

Figure 8.10

ⓗ Writing ionic equations

- When an ionic substance dissolves in water, the positive and negative ions separate and do not interact with each other.
- When solutions containing ions react in solution, some of the ions do not react. We can write an ionic equation that leaves out any ions that are unchanged during the reaction.
- For example, when chlorine displaces bromine from sodium bromide solution:

Full equation: $Cl_2 + 2NaBr \rightarrow 2NaCl + Br_2$

Showing all the ions separately: $Cl_2 + 2Na^+ + 2Br^- \rightarrow 2Na^+ + 2Cl^- + Br_2$

Cancelling out ions that do not change: $Cl_2 + 2Na^+ + 2Br^- \rightarrow 2Na^+ + 2Cl^- + Br_2$

Final ionic equation: $Cl_2 + 2Br^- \rightarrow 2Cl^- + Br_2$

> **Typical mistake**
>
> Some students write about the electron in the outer shell when explaining the trend in reactivity down Group 7. The atoms gain an electron when they react and this comes from outside the atom – it does not come from its own outer shell.

> **Ionic equation** an equation that leaves out any ions that do not change during a reaction

Now test yourself

TESTED ☐

18 Why do the halogens have low melting and boiling points?
19 When an aqueous solution of bromine is mixed with an aqueous solution of potassium iodide, the mixture darkens to brown.
   (a) Write a word equation for this reaction.
   (b) Write an ionic equation for this reaction.
   (c) Is bromine or iodine more reactive?
   (d) Explain your answer to (c).

Answers online

Development of the periodic table

REVISED ☐

- Before the discovery of protons, neutrons and electrons, scientists had tried to classify the elements by putting them into order of atomic mass.
- There were many problems with these early attempts with elements out of place (for example, in one attempt, the unreactive metal copper was in the same group as the reactive metals sodium and potassium).
- Dimitri Mendeleev made a big breakthrough when he produced a table where the elements were mainly in mass order, but:
  - he left gaps for elements he predicted would be discovered (and predicted what their properties would be)
  - he slightly changed the order of some elements out of mass order so they were in groups with elements with similar properties.
- Over the next few years, some of the elements Mendeleev predicted were discovered and their properties matched his predictions.
- Years later, when protons were discovered, scientists realised that Mendeleev had placed the elements in order of atomic number.

Mixtures

Comparing a mixture to a compound

Table 8.11

	mixture	compound
description	two or more substances that are mixed together and not chemically combined	a substance made from two or more elements chemically bonded together
properties	each substance in the mixture has its own properties	the compound has its own unique properties that are different from the elements it is made from
proportions	the substances can be mixed in any proportion	the elements are combined in a fixed proportion
separation	easy to separate the substances by physical methods as they are not chemically bonded together	cannot separate the substances by physical methods as they are chemically bonded together (the compound could only be decomposed back into elements by a chemical reaction)

Separating mixtures

Table 8.12

method	filtration	evaporation / crystallisation	chromatography
what is separated	insoluble solid and a liquid	soluble solid from a solution	mixture of soluble solids
details	the liquid goes through the filter paper; the solid does not go through the filter paper filter funnel / filter paper / solid (residue) collects on filter paper / conical flask / liquid (filtrate) collects in flask	evaporation: the solvent evaporates leaving the solid evaporating dish / heat crystallisation: some of the solvent is evaporated and then crystals of the solid form as the solution cools	the substances dissolve in the solvent and move up the paper at different speeds paper held up by splint or rod / chromatography paper / beaker / pencil start line / solvent / Y A B C
example	sand and water	salt from a solution of salt water	mixture of dyes

Table 8.12 *continued*

method	simple distillation	fractional distillation	separating funnel
what is separated	solvent from a solution	miscible liquids (i.e. liquids that mix together)	immiscible liquids (i.e. liquids that do not mix together)
details	the solvent boils off leaving the solid behind; the vaporized solvent cools and is condensed back into a liquid	the liquids have different boiling points and so boil off separately	the liquids form two layers as they do not mix together
example	water from a solution of sea water	ethanol from a mixture of water and ethanol	oil from a mixture of oil and water

Now test yourself

TESTED

20 What method would you use to separate each of the following mixtures?
  (a) ethanol from a solution of potassium hydroxide in ethanol
  (b) ethanol from a mixture of ethanol and octane (they are miscible liquids)
  (c) potassium bromide from a solution of potassium bromide in water
  (d) water from a suspension of barium sulfate in water (barium sulfate is insoluble in water)

Answers online

Summary

- Atoms contain protons and neutrons in a tiny nucleus surrounded by electrons in energy levels (shells).
- The atomic number is the number of protons in an atom.
- The mass number is the number of protons plus the number of neutrons in an atom.
- Atoms are neutral and so the number of protons and electrons is equal.
- Ions are charged particles which have a different number of protons and electrons.
- Electrons fill the lowest energy levels (shells) available, with up to 2 electrons in the 1st energy level, up to 8 electrons in the 2nd energy level, the next 8 electrons in the 3rd energy level and the next 2 electrons in the 4th energy level.

- Isotopes are atoms with the same number of protons but a different number of neutrons.
- Relative atomic mass of an element is the average mass of the isotopes.
- Ideas about the structure of atoms have changed as evidence has been discovered that did not fit with the existing model.
- Elements react with each other to form compounds. When metals react with non-metals, electrons are transferred and an ionic compound is formed. When non-metals react with non-metals, electrons are shared to form molecular compounds.
- Elements are listed in the periodic table and are in atomic number order.

- Elements in the same group of the periodic table have similar properties because they have the same number of electrons in the outer shell.
- Group 0 elements (noble gases) are unreactive gases because they have a stable electron structure.
- Group 1 elements (alkali metals):
  - are very reactive metals,
  - all have one electron in their outer shell, which they lose when they react to form 1+ ions in ionic compounds,
  - get more reactive going down the group as the outer shell electron is easier to lose as the atoms get bigger and so the outer electron has a weaker attraction to the nucleus.
- Group 7 elements (the halogens):
  - are very reactive non-metals,
  - all have seven electrons in their outer shell and gain one more when they react either by forming covalent bonds or by forming 1– ions,
  - get less reactive going down the group as it becomes harder to gain one electron as the atoms get bigger and so the electron gained has a weaker attraction to the nucleus,
  - can displace a less reactive halogen from its compounds.
- The modern periodic table is based on Mendeleev's table where he:
  - put most of the elements in mass order but changed it around to fit the properties,
  - predicted the existence and properties of new elements that he left gaps for – these elements were discovered and fitted his predictions.
- Mixtures contain substances that are not chemically bonded to each other.
- Mixtures can be separated by:
  - filtration to separate an insoluble solid that is mixed with a liquid,
  - evaporation or crystallisation to separate a solid from a solution,
  - simple distillation to separate a solvent from a solution,
  - fractional distillation to separate a mixture of miscible solvents,
  - separating funnel to separate immiscible liquids,
  - chromatography to separate a mixture of solids.

Exam practice

1 The element boron consists of the two isotopes $^{10}_{5}B$ and $^{11}_{5}B$.
  (a) Describe the similarities and differences between these atoms in terms of their number of protons, neutrons and electrons. [3 marks]
  (b) Give the electron structure of $^{11}_{5}B$. [1 mark]
  (c) Explain why boron is in group 3 of the periodic table. [1 mark]
  (d) 20% of boron atoms are $^{10}_{5}B$, with the other 80% being $^{11}_{5}B$. Calculate the relative atomic mass of boron. Give your answer to 3 significant figures. [3 marks]
2 Argon (atomic number 18) is very unreactive while potassium (atomic number 19) is very reactive. Explain this difference in reactivity. [4 marks]
3 Potassium is in Group 1 and reacts vigorously with water.
  (a) Write an equation for the reaction of potassium with water. [2 marks]
  (b) Give the charge and electron structure of the ion formed from potassium atoms when they react. [2 marks]
  (c) What colour would universal indicator turn if added to the water after the reaction? Explain your answer. [3 marks]
  (d) Explain why potassium reacts more vigorously with water than sodium. [3 marks]
4 Iodine is in Group 7.
  (a) Which of the following is the correct formula of iodine molecules? **A** I   **B** I_2 **C** I⁻   **D** I_2^- [1 mark]
  (b) Explain why a solution of iodine in water does not react with a solution of potassium bromide. [1 mark]
5 What contribution did Dimitri Mendeleev make to the development of the periodic table? Explain why his ideas were accepted and what order he actually placed the elements in without knowing. [6 marks]
6 Describe how you would separate water from salt water. [4 marks]

Answers and quick quiz 8 online

ONLINE

9 Bonding, structure and the properties of matter

The key to understanding chemistry is to understand the structure of a substance at the particle level in terms of what the particles are and how they are bonded or attracted to each other.

Types of bond

- There are three types of bond which are described in the table.

Table 9.1

name	ionic	covalent	metallic
description	the attraction between positive and negative ions	atoms joined together by sharing pairs of electrons	the attraction between the positive nuclei of metals atoms and delocalised electrons
types of substances with this bonding	ionic substances	molecular substances; giant covalent substances	metallic elements and alloys

Ionic substances

The structure of ionic substances   REVISED

- Ions are electrically charged particles that contain a different number of protons and electrons.
- Most compounds that are made from a combination of metals and non-metals are made up of ions (e.g. NaCl is made from the metal sodium and the non-metal chlorine).
- They are all solids at room temperature and the positive and negative ions are arranged in a **giant lattice**. A giant lattice contains a massive number of particles that continues in all directions throughout the substance.
- There is a strong attraction between the positive and negative ions as opposite charges attract.

> **Typical mistake**
>
> The particles in an ionic substance are ions. There are no atoms or molecules in the substance.

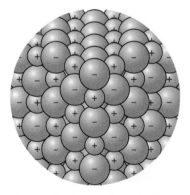

Figure 9.1 A small part of the giant lattice of positive and negative ions in sodium chloride (NaCl)

Properties of ionic substances   REVISED

- **Melting and boiling points** – high because there is a strong attraction between the positive and negative ions.
- **Electrical conductivity as solids** – do not conduct electricity as the ions cannot move.
- **Electrical conductivity as liquids or when dissolved** – do conduct electricity as the ions can move through the substance.

Comparing ways to represent the structure of ionic compounds

- The structure of an ionic compound can be represented in several ways each of which has its advantages and disadvantages.

Table 9.2

diagram ⊖ Cl⁻ ion ⊕ Na⁺ ion	dot and cross diagram	2D space-filling structure	3D space-filling structure	ball and stick structure
	$\left[\text{Na}\right]^+$ (2,8) $\left[\stackrel{\times\times}{\stackrel{\times}{\underset{\times\times}{\text{Cl}}}}\times\right]^-$ (2,8,8)			the ions are shown with gaps between them lines are drawn between the ions to show how they are arranged
advantages of this representation	shows the electron structure of the ions	very easy to draw	gives very good representation of how the ions are packed together	helps to show how the ions are arranged relative to each other
disadvantages of this representation	can give the impression that the structure is made of pairs of ions rather than being a continuous structure containing a massive number of ions	can give the impression that the structure is limited to a few ions rather than being a continuous structure with a massive number of ions		
		only shows the structure in 2D		may make you think there are covalent bonds between the ions (there are **NO** covalent bonds in an ionic lattice) may make you think the ions are a long way apart (but they are packed close together)

The formation of ionic compounds

- Ionic compounds are formed when metals react with non-metals.
- Metal atoms lose electrons to form positive ions. Non-metal atoms gain these electrons to form negative ions. In each case, the ions formed have stable noble gas (Group 0) electron structures.
- For example, sodium atoms react with chlorine atoms to form the ionic compound sodium chloride. This can be shown in several ways.

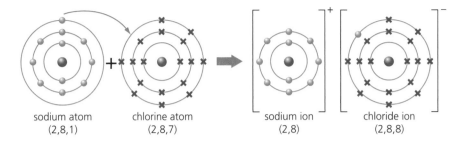

sodium atom (2,8,1) chlorine atom (2,8,7) sodium ion (2,8) chloride ion (2,8,8)

Only showing the outer shell electrons:

Na + Cl ⟶ [Na]⁺ + [Cl]⁻

(2,8,1)   (2,8,7)   (2,8)   (2,8,8)

Figure 9.2

Exam tip

When an atom loses electrons, it forms an ion with a positive charge. When an atom gains electrons, it forms an ion with a negative charge.

- The table shows how atoms in some groups of the periodic table react.

Table 9.3

	Group 1	Group 2	Group 6	Group 7
number of electrons in outer shell	1	2	6	7
what happens when they react	lose 1 electron	lose 2 electrons	gain 2 electrons	gain 1 electron
charge on ion formed	1+	2+	2–	1–
examples	Na^+, K^+	Mg^{2+}, Ca^{2+}	O^{2-}, S^{2-}	Cl^-, Br^-

The formula of ionic compounds

REVISED

- The formula represents the ratio of the ions in the lattice. For example, the formula of aluminium oxide is Al_2O_3, which means that there are two aluminium ions for every three oxide ions throughout the lattice structure.
- Ionic compounds have no overall electric charge. Therefore, the total number of positive charges must equal the total number of negative charges. This allows us to work out the formula of an ionic compound.
- For example, potassium oxide contains K^+ and O^{2-} ions. For the charges to balance, there must be two K^+ ions for every one O^{2-} ion, and so the formula is K_2O.

Table 9.4

name of compound	ions in compound	ratio of ions so that the charges balance		formula
potassium oxide	K^+ and O^{2-}	K^+	O^{2-}	K_2O
		K^+		
		2+	2–	
sodium chloride	Na^+ and Cl^-	Na^+	Cl^-	NaCl
		1+	1–	
aluminium oxide	Al^{3+} and O^{2-}	Al^{3+}	O^{2-}	Al_2O_3
		Al^{3+}	O^{2-}	
			O^{2-}	
		6+	6–	

➜

Table 9.4 *continued*

name of compound	ions in compound	ratio of ions so that the charges balance		formula
calcium nitrate	Ca^{2+} and NO_3^-	Ca^{2+}	NO_3^-	$Ca(NO_3)_2$
			NO_3^-	
		2+	2−	
iron(III) hydroxide	Fe^{3+} and OH^-	Fe^{3+}	OH^-	$Fe(OH)_3$
			OH^-	
			OH^-	
		3+	3−	

Exam tip

- You need to be able to work out the charge of ions of elements in Groups 1, 2, 6 and 7. You will be given the charge on other ions.
- Some elements form ions with different charges. With their ions, the charge on the ion is given as a roman numeral. For example, iron(III) ions are Fe^{3+}.

Typical mistake

Some ions contain different elements combined (e.g. NO_3^-, OH^-, CO_3^{2-}, SO_4^{2-}, NH_4^+). If there is more than one of these in the formula, then it must be written in a bracket (even when it is OH^-).

Now test yourself

TESTED

1 Calcium oxide is an ionic compound.
   (a) Explain why calcium oxide has a high melting point.
   (b) Explain why calcium oxide conducts electricity when molten but not as a solid.
   (c) Give the formula and charge of the ions in calcium oxide.
2 Give the formula of each of these ionic compounds (sulfate ions = SO_4^{2-}, hydroxide ions = OH^-).
   (a) magnesium fluoride
   (b) copper(II) sulfide
   (c) potassium sulfate
   (d) calcium hydroxide

Answers online

Molecular (covalent) substances

The structure of molecular substances

REVISED

- Molecules are particles made from two or more atoms joined by covalent bonds.
- The covalent bonds holding the atoms together within each molecule are very strong. However, the molecules are not bonded to each other and there are only weak forces between molecules.
- Some non-metal elements are made from molecules (e.g. H_2, N_2, O_2, F_2, Cl_2, Br_2, I_2 and S_8).
- Most compounds made from non-metals combined with other non-metals are made from molecules (e.g. water – a compound made from the non-metals hydrogen and oxygen).
- The formula of a molecular substance gives the number of atoms of each element in one molecule. For example, the formula of water is H_2O, which means that there are two H atoms and one O atom in each molecule.

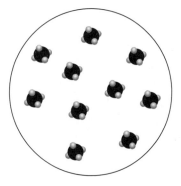

Figure 9.3 Molecules in methane (CH_4)

Properties of molecular substances

- **Melting and boiling points** – low because there are weak forces between the molecules; these forces increase as molecules get bigger and so larger molecules have higher melting and boiling points; many molecular substances are gases or liquids at room temperature.
- **Electrical conductivity** – do not conduct electricity as they do not contain any delocalised electrons or ions.

> **Typical mistake**
>
> Covalent bonds are strong. When a molecular substance melts or boils, covalent bonds do not break. For example, when water boils to form steam, the covalent bonds in the H_2O molecules do not break.

Drawing molecules

- When atoms join together to form molecules they share electrons in order to obtain stable noble gas electron structures.
- A **covalent bond** is formed when two shared electrons join atoms together. Atoms can be joined by single (two electrons), double (four electrons) or triple (six electrons) covalent bonds.
- The number of covalent bonds an atom forms equals the number of electrons it needs to obtain a noble gas electron structure.

Table 9.5

	H	Group 4 atoms (e.g. C, Si)	Group 5 atoms (e.g. N, P)	Group 6 atoms (e.g. O, S)	Group 7 atoms (e.g. F, Cl)
number of electrons in outer shell	1	4	5	6	7
number of electrons needed to obtain a noble gas electron structure	1	4	3	2	1
number of covalent bonds formed	1	4	3	2	1

- Molecules can be represented in several ways, each of which has its advantages and disadvantages.

Table 9.6

dot and cross diagram showing all electrons and shell circles	dot and cross diagram showing only outer shell electrons and shell circles	dot and cross diagram showing only outer shell electrons	stick diagram	ball and stick diagram	space-filling diagram
note that the • and × represent electrons that came from different atoms			each stick (or line) represents one covalent bond (i.e. 2 shared electrons)		a good representation of a molecule showing how atoms merge into each other, but the covalent bonds are not visible

- When drawing stick diagrams, each atom makes the number of bonds it needs to obtain a noble gas electron structure.
- When drawing dot-cross diagrams, each covalent bond is made up of two electrons. Any other electrons that are not used in making covalent bonds are shown in non-bonding pairs (often called lone pairs).

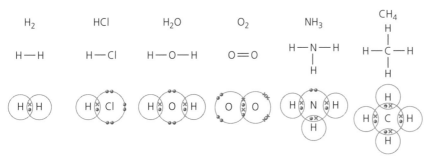

Figure 9.4

Polymers

- **Polymers** (e.g. polythene, PVC) are long chain molecules made from joining lots of short molecules together.
- Each polymer chain is a very long molecule with atoms joined by covalent bonds.
- Molecules that are very long are called **macromolecules**.
- In a **thermosoftening polymer**, the polymer chains are not joined together. This means that they will soften and melt on heating. They are solids at room temperature because the forces between the molecules are relatively strong as they are big molecules.

> **Polymer** long chain molecule made from joining lots of short molecules together
>
> **Macromolecule** a very long molecule

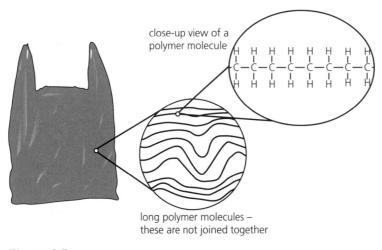

close-up view of a polymer molecule

long polymer molecules – these are not joined together

Figure 9.5

Now test yourself

3 Methane (CH_4) is a gas at room temperature. The atoms within the molecule are joined by covalent bonds.
  (a) What is a covalent bond?
  (b) Explain why methane has a low boiling point.
4 Draw a stick and a dot-cross diagram for:
  (a) F_2
  (b) PH_3
  (c) CO_2
5 Polystyrene is a thermosoftening polymer. Explain why it softens and melts when heated.

Answers online

Giant covalent substances

The structure of giant covalent substances REVISED

- There are a few substances that have atoms joined by covalent bonds in one giant continuous network.
- Examples include three forms of carbon (diamond, graphite, graphene), silicon and silicon dioxide.

Typical mistake

Giant covalent substances are **not** big molecules, they are a continuous network of atoms linked by covalent bonds.

Properties of giant covalent substances REVISED

- **Melting and boiling points** – very high because covalent bonds must be broken.
- **Electrical conductivity** – most do not conduct electricity as they have no delocalised electrons; graphite and graphene do conduct electricity as they contain some delocalised electrons.

Now test yourself TESTED

6 Diamond has a very high melting point at 3550°C. Explain why it is so high.

Answers online

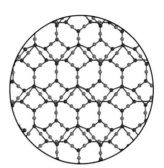

Figure 9.6 **A small part of the giant lattice of atoms joined by covalent bonds in silicon dioxide (SiO$_2$)**

Metallic substances

The structure of metallic substances REVISED

- Metals are metallic substances.
- They consist of a giant lattice of atoms, but the outer shell electrons from each atom are delocalised (which means that they are free to move throughout the metal).
- There is a strong attraction between the positive nucleus of the metal atoms and the negative delocalised electrons (this is called **metallic bonding**).

Properties of metallic substances REVISED

- **Melting and boiling points** – high because the attraction between the positive nucleus of the metal atoms and the negative delocalised electrons is strong.
- **Electrical and thermal conductivity** – they conduct because the outer shell electrons are delocalised and can carry the charge through the metal.
- **Malleability** – they are malleable (can be hammered into shape) because the atoms can slide over each other while maintaining the metallic bonding.

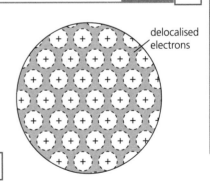

Figure 9.7 **A small part of the giant lattice of atoms in a cloud of delocalised outer shell electrons**

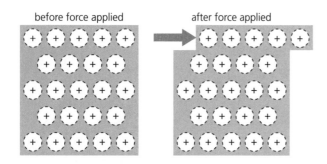

Figure 9.8 Metals are malleable

Alloys

- Pure metals are malleable which can make them too soft for many uses.
- An **alloy** is a mixture of a metal with small amounts of other metals or carbon added (e.g. steels are alloys of iron).
- Alloys are harder than pure metals because they contain some atoms that are a different size and so it is more difficult for atoms to slide past each other.

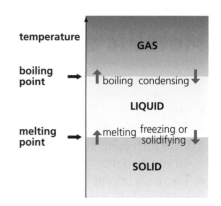

Figure 9.9 An alloy

Now test yourself

7 Iron is a metal with a melting point of 1538°C.
   (a) Explain why iron has a high melting point.
   (b) Explain why iron conducts electricity.
   (c) Steels are alloys of iron. Explain why steels are harder than iron.

Answers online

States of matter

- The three states of matter are solid, liquid and gas. The state that a substance is in depends on the temperature (Figure 9.10).
- The three states of matter can be represented by a very simple model where particles are represented by small solid spheres (Figure 9.11).

traditional model of particles in solids, liquids and gas

Solid Liquid Gas

Figure 9.11

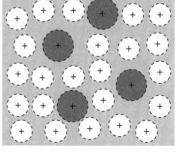

Figure 9.10 Changes of state

H ● However, this model is limited because
   ○ it does not show the forces or bonds between the particles
   ○ it does show what the particles are like (e.g. they could be molecules or ions).
- The stronger the forces or bonds between the particles, the higher the melting and boiling points. For example,
   ○ in a molecular substance, there are only weak forces between the molecules – this means that it only takes a small amount of energy to separate the molecules and so it has a low melting and boiling point.

○ in an ionic substance, there are strong attractions between the
positive and negative ions – this means that it takes a lot of energy to
separate the particles and so it has a high melting and boiling point.

Now test yourself

TESTED ▢

8 The melting and boiling points of some substances are shown.

substance	melting point (°C)	boiling point (°C)
A	747	1390
B	6	81
C	114	184
D	−78	−15

(a) What state is each substance in at 20°C?
(b) Which substance is a liquid at 150°C?
(c) Which substance condenses as it is cooled from 100°C to 20°C?
(d) Which substance requires the most energy to melt?

Answers online

Different forms of carbon

● The element carbon exists in several different forms.

Table 9.7

	diamond	graphite	buckminsterfullerene	graphene
formula	C	C	C_{60}	C
structure	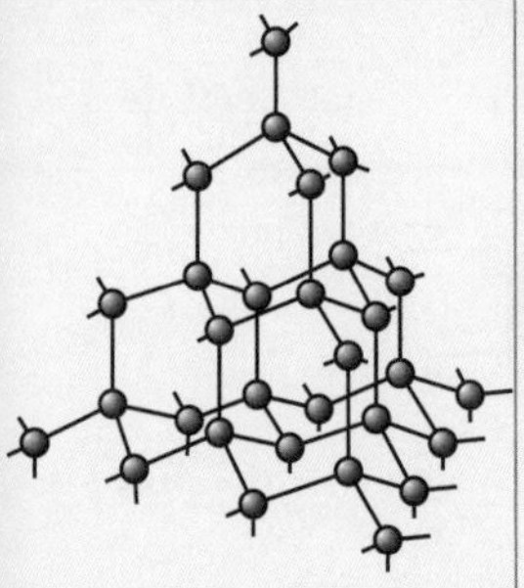	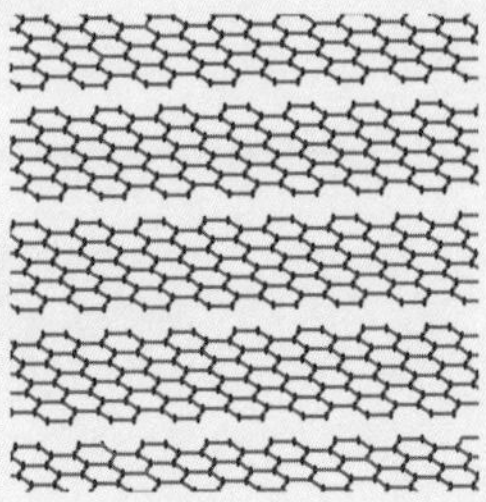	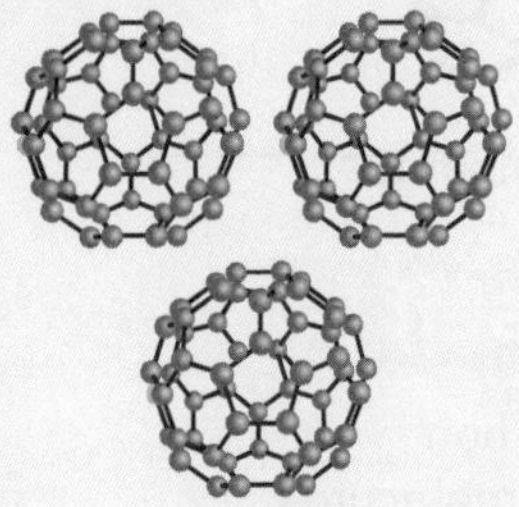	
structure	giant covalent	giant covalent	simple molecular	giant covalent
description	A lattice of atoms joined by covalent bonds. Each C atom makes 4 bonds.	A lattice of atoms joined by covalent bonds. Each C atom makes 3 bonds. The atoms are in flat layers but there are only weak forces between these layers. The layers can move past each other. One electron from each atom is delocalised and can move along the layers.	This is made up of many molecules. Each molecule contains 60 carbon atoms. Each C atom makes 3 bonds. One electron from each atom is free to move within the molecule.	This is a single layer of graphite. It has a lattice of atoms joined by covalent bonds. Each C atom makes 3 bonds. One electron from each atom is delocalised and can move across the surface.

→

Table 9.7 *continued*

	diamond	graphite	buckminsterfullerene	graphene
melting and boiling points	very high (due to the need to break strong covalent bonds)	very high (due to the need to break strong covalent bonds)	not so high (no covalent bonds are broken)	very high (due to the need to break strong covalent bonds)
electrical conductivity	insulator (no delocalised electrons)	conductor (has delocalised electrons that can move along layers)	insulator (has delocalised electrons but they cannot move from one molecule to another)	conductor (has delocalised electrons that can move across the surface)
strength	very hard and strong	soft and brittle	soft and brittle	very strong
uses	cutting tools	electrodes, pencils	drug delivery, lubricants	electronic components

- Buckminsterfullerene (C_{60}) is the most common fullerene. **Fullerenes** are a family of molecules containing carbon atoms in hollow shapes made up of rings of 5, 6 or 7 carbon atoms.
- Carbon nanotubes are related to fullerenes. They are very long, tubes with a high length to diameter ratio. They are very strong and are good conductors. They have many uses, including in sports equipment such as tennis racquets, to make them light but strong.

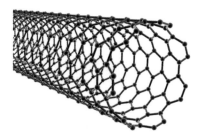

Figure 9.12 A carbon nanotube

Now test yourself

TESTED ☐

9   Name a form of carbon
(a) that is made of molecules
(b) that is used in drill tips
(c) in which each C atom forms 4 covalent bonds.
10 Explain why
(a) diamond has a high melting point
(b) graphene is an electrical conductor
(c) graphite is soft and brittle.

Answers online

Summary

- There are three different types of bonding: ionic, covalent and metallic.
- Substances usually have one of the following five types of structure.

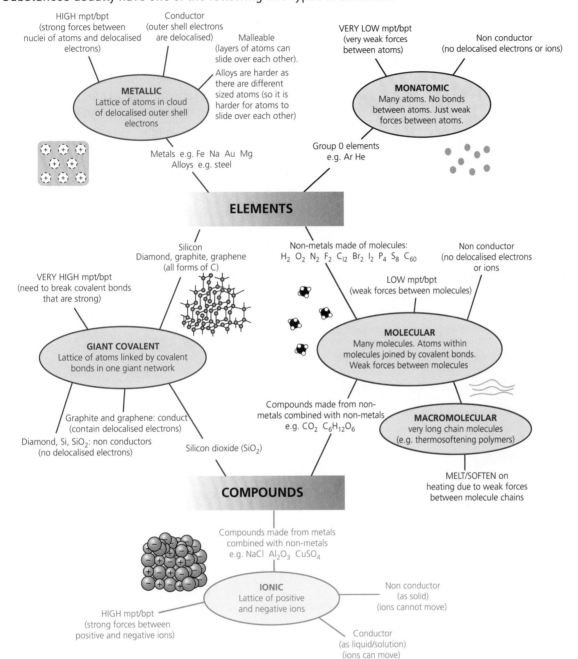

HIGH mpt/bpt
(strong forces between nuclei of atoms and delocalised electrons)

Conductor
(outer shell electrons are delocalised)

Malleable
(layers of atoms can slide over each other).

Alloys are harder as there are different sized atoms (so it is harder for atoms to slide over each other)

VERY LOW mpt/bpt
(very weak forces between atoms)

Non conductor
(no delocalised electrons or ions)

METALLIC
Lattice of atoms in cloud of delocalised outer shell electrons

MONATOMIC
Many atoms. No bonds between atoms. Just weak forces between atoms.

Metals e.g. Fe Na Au Mg
Alloys e.g. steel

Group 0 elements
e.g. Ar He

ELEMENTS

Silicon
Diamond, graphite, graphene (all forms of C)

Non-metals made of molecules:
H_2 O_2 N_2 F_2 Cl_2 Br_2 I_2 P_4 S_8 C_{60}

Non conductor
(no delocalised electrons or ions)

VERY HIGH mpt/bpt
(need to break covalent bonds that are strong)

LOW mpt/bpt
(weak forces between molecules)

GIANT COVALENT
Lattice of atoms linked by covalent bonds in one giant network

MOLECULAR
Many molecules. Atoms within molecules joined by covalent bonds. Weak forces between molecules

Graphite and graphene: conduct (contain delocalised electrons)

Diamond, Si, SiO_2: non conductors (no delocalised electrons)

Silicon dioxide (SiO_2)

Compounds made from non-metals combined with non-metals e.g. CO_2 $C_6H_{12}O_6$

MACROMOLECULAR
very long chain molecules (e.g. thermosoftening polymers)

COMPOUNDS

MELT/SOFTEN on heating due to weak forces between molecule chains

Compounds made from metals combined with non-metals e.g. NaCl Al_2O_3 $CuSO_4$

IONIC
Lattice of positive and negative ions

Non conductor
(as solid)
(ions cannot move)

HIGH mpt/bpt
(strong forces between positive and negative ions)

Conductor
(as liquid/solution)
(ions can move)

- Carbon exists in several forms, including diamond, graphite, graphene and buckminsterfullerene (an example of a fullerene).

Exam practice

1 Describe what happens in terms of electrons when magnesium atoms react with fluorine atoms. [4 marks]

2 What is the charge on a calcium ion?
   A 1+   B 2+   C 1−   D 2− [1 mark]

3 Which of the following is the correct formula of sodium oxide? A = NaO, B = NaO_2, C = NaO_3, D = Na_2O [1 mark]

→

4 Match the following substances to their structure type. [3 marks]

magnesium oxide (MgO)		giant covalent
bromine (Br$_2$)		ionic
graphene (C)		metallic
		molecular

5 Which one of the following substances is an electrical insulator as a solid?
A graphene   B graphite   C iron   D sodium chloride   [1 mark]

6 Diamond and graphite are both forms of carbon. Describe some similarities and differences in their physical properties and explain reasons for these in terms of their structure and bonding. [6 marks]

7 Pure gold is a soft metal that is an excellent electrical conductor. Explain why pure gold is:
(a) soft [2 marks]
(b) an electrical conductor. [3 marks]

8 The structure of a molecule of water is shown. Draw a diagram showing the outer shell electrons of the atoms in this molecule, using crosses to represents those from hydrogen and dots for those from oxygen. [2 marks]

H—O—H

(H) 9 The diagram below could be used as a very simple model of the structure of solid ice. Describe some limitations with this model. [6 marks]

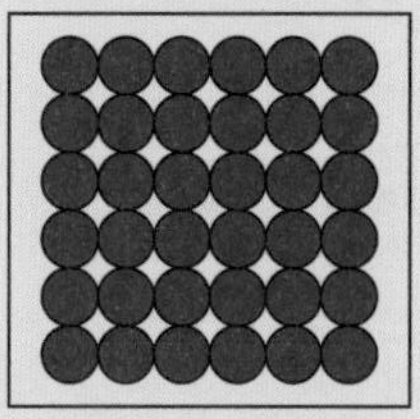

Answers and quick quiz 9 online

ONLINE

10 Quantitative chemistry

Conservation of mass

- The total mass of the reactants must always equal the total mass of the products in any chemical reaction. This is called the law of **conservation of mass**. This happens because the reactants and products are made of exactly the same atoms.

Thermal decomposition reactions

- A thermal decomposition reaction is one where heat causes a substance to break down into simpler substances. One or more of the products may escape as gases which may appear to break the law, but the total mass of all the products will equal the total mass of the reactants.
- For example, if 100 g of calcium carbonate is heated, only 56 g of calcium oxide remains. However, the other 44 g has escaped as carbon dioxide gas, so there are still 100 g of products and mass is conserved.

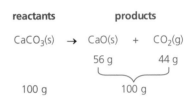

Reactions of metals with oxygen

- When metals react with oxygen, the mass of the product is greater than the mass of the original metal. This may appear to break the law, but when the mass of the oxygen from the air is added in, it fits the law.
- For example, if 48 g of magnesium is heated, 80 g of magnesium oxide is formed. However, the other 32 g has come from oxygen from the air.

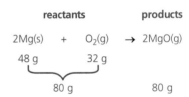

Now test yourself

1 Copper carbonate decomposes when it is heated. When 6.2 g of copper carbonate is heated, some carbon dioxide gas and 4.0 g of copper oxide is formed. Use the law of conservation of mass to calculate the mass of carbon dioxide formed.

Answers online

Relative formula mass

- The **relative formula mass** (M_r) of a substance is the sum of the relative atomic masses (A_r) of all the atoms shown in the formula.

> **Example**
>
> | H_2O | $M_r = 2(1) + 16 = 18$ |
> | Cl_2 | $M_r = 2(35.5) = 71$ |
> | $Ca(NO_3)_2$ | $M_r = 40 + 2(14) + 6(16) = 164$ |

> **Exam tip**
>
> You may be told the relative atomic mass (A_r) of each atom in a question, but they are also shown on the periodic table on the data sheet. There are two numbers in each box on the table – the A_r is the bigger of the two numbers.

TESTED ☐

Now test yourself

2 What is the relative formula mass (M_r) of the following?
   (a) CaO   (A_r Ca = 40, O = 16)
   (b) O_2 (A_r O = 16)
   (c) H_2SO_4 (A_r H = 1, S = 32, O = 16)
   (d) $Al_2(SO_4)_3$ (A_r Al = 27, S = 32, O = 16)

Answers online

> **Exam tip**
>
> Relative formula mass (M_r) can also be called relative molecular mass, molecular mass, formula mass or just M_r.

The mole

REVISED ☐

- Amounts of chemicals are often measured in moles.
- One mole of particles of any substance contains 6.02×10^{23} (602 000 000 000 000 000 000 000) particles (this number is known as the Avogadro constant).
- The mass of one mole of any substance is the M_r in grams, e.g. the mass of one mole of H_2O ($M_r = 18$) = 18 g, the mass of one mole of Cl_2 ($M_r = 71$) = 71 g
- Mass, moles and M_r are linked by this equation:

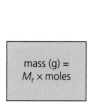

mass (g) = M_r × moles

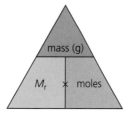

To use the formula triangle, cover up the quantity you are trying to calculate. This then shows you the equation to use.

Figure 10.1

> **Exam tip**
>
>
>
> **Figure 10.2**
>
> Thinking of 'Mr Moles' with a mass on his head may help you to remember the formula triangle.

Example

How many moles are there in 90 g of water (H_2O)? (A_r H = 1, O = 16)

$$\text{moles} = \frac{\text{mass}}{M_r} = \frac{90}{18} = 5 \text{ mol}$$

What is the mass of 3 moles of oxygen (O_2)? (A_r O = 16)

$$\text{mass} = M_r \times \text{moles} = 32 \times 3 = 96 \text{ g}$$

- When working with this equation, the mass must be in grams. Some conversion factors are shown.

tonnes $\xrightarrow{\times\,1\,000\,000}$ g   e.g. 5 tonnes = 5 000 000 g

kg $\xrightarrow{\times\,1000\,g}$ g   e.g. 2 kg = 2000 g

mg $\xrightarrow{\div\,1000\,g}$ g   e.g. 15 mg = 0.015 g

Example

How many moles are there in 36 kg of water (H_2O)? (A_r H = 1, O = 16)

$$\text{moles} = \frac{\text{mass}}{M_r} = \frac{36\,000}{18} = 2000 \text{ mol}$$

How many moles are there in 14 mg of calcium oxide (CaO)? (A_r Ca = 40, O = 16)

$$\text{moles} = \frac{\text{mass}}{M_r} = \frac{0.014}{56} = 0.00025 \text{ mol (i.e. } 2.5 \times 10^{-4}\text{)}$$

> **Typical mistake**
>
> The unit abbreviation for moles is mol (not m which is metres!).

Significant figures

- You may be asked to give an answer to a specific number of significant figures (sf). The table illustrates different numbers of significant figures.

Table 10.1

number	2 sf	3 sf	4 sf
432821	430000	433000	432800
0.0078932	0.0079	0.00789	0.007893

Exam tip

With decimal numbers, the first significant figure is the first non-zero number after the decimal point (see example).

Now test yourself

3 How many moles are there in each of the following?
   (a) 36 g of water (H_2O) (A_r H = 1, O = 16)
   (b) 11 tonnes of carbon dioxide (CO_2) (A_r C = 12, O = 16)
   (c) 30 mg of sodium hydroxide (NaOH) (A_r Na = 23, O = 16, H = 1)
4 What is the mass of each of the following?
   (a) 2.5 mol of oxygen (O_2) (A_r O = 16)
   (b) 0.30 mol of magnesium bromide ($MgBr_2$) (A_r Mg = 24, Br = 80)

Answers online

Molar ratios in equations

- A balanced equation shows the ratio in moles in which substances react. For example:

N_2	+	$3H_2$	→	$2NH_3$
1 mole of N_2		3 moles of H_2		2 moles of NH_3
10 moles of N_2		30 moles of H_2		20 moles of NH_3
20 moles of N_2	reacts with	60 moles of H_2	to make	40 moles of NH_3
0.1 moles of N_2		0.3 moles of H_2		0.2 moles of NH_3

Now test yourself

5 In the following reaction: $2H_2 + O_2 \rightarrow 2H_2O$
   (a) How many moles of hydrogen react with 10 moles of oxygen?
   (b) How many moles of oxygen react with 1 mole of hydrogen?
   (c) How many moles of water are formed when 5 moles of oxygen react with lots of hydrogen?
   (d) How many moles of water are formed when 0.4 moles of hydrogen react with lots of oxygen?

Answers online

Reacting mass calculations

- By using moles, we can calculate the mass of chemicals that are involved in chemical reactions.
- In these calculations:
   (a) calculate the number of moles of the substance whose mass is given (using moles = $\frac{mass}{M_r}$)
   (b) calculate the number of moles of the substance the question is about (using ratios from the balanced equation)
   (c) calculate the mass of the substance the question asks about (using mass = $M_r \times$ moles).

Example

What mass of magnesium reacts with 8 g of oxygen? (A_r Mg = 24, O = 16)

$$2Mg + O_2 \rightarrow 2MgO$$

(a) moles of $O_2 = \dfrac{mass}{M_r} = \dfrac{8}{32} = 0.25$ mol

(b) moles of Mg = 2 × 0.25 = 0.50 mol (as the equation shows that 2 moles of Mg react with 1 mole of O_2)

(c) mass of Mg = M_r × moles = 24 × 0.50 = **12 g**

Example

What mass of ammonia is formed when 12 g of hydrogen reacts with nitrogen? (A_r H =1, N = 14)

$$N_2 + 3H_2 \rightarrow 2NH_3$$

(a) moles of $H_2 = \dfrac{mass}{M_r} = \dfrac{12}{2} = 6$ mol

(b) moles of $NH_3 = 6 \times \dfrac{2}{3} = 4$ mol (as the equation shows that 3 moles of H_2 makes 2 moles of NH_3)

(c) mass of NH_3 = M_r × moles = 17 × 4 = **68 g**

> **Exam tip**
>
> Always look at the balancing numbers in the equation to see if the number of moles of the substance you are calculating should be bigger or smaller than the moles you start with.

Now test yourself

TESTED

6 What mass of calcium oxide is formed when 60 g of calcium carbonate decomposes? (A_r Ca = 40, C = 12, O = 16)

$$CaCO_3 \rightarrow CaO + CO_2$$

7 What mass of chlorine reacts with 5.6 g of iron? (A_r Fe = 56, Cl = 35.5)

$$2Fe + 3Cl_2 \rightarrow 2FeCl_3$$

8 What mass of aluminium is made from 1.53 kg of aluminium oxide? (A_r Al = 27, O = 16)

$$2Al_2O_3 \rightarrow 4Al + 3O_2$$

Answers online

Using moles to balance equations

- The balancing numbers in an equation can be calculated from the masses of reactants and products that are involved in a reaction.
- In these calculations:
  (a) calculate the number of moles of each substance (using $moles = \dfrac{mass}{M_r}$)
  (b) divide each number of moles by the smallest one to give the simplest whole number ratio
  (c) if this does not give a whole number ratio, you may need to times by a number such as 2 or 3 until you get a whole number ratio.

> **Typical mistake**
>
> Some students go wrong in questions like this because they round numbers too much and too soon. Never use numbers to less than 3 significant figures during these questions (although your final answers at the end should be written as integers (whole numbers)).

Example

20 g of calcium (Ca) reacts with 8 g of oxygen (O_2) to form 28 g of calcium oxide (CaO). Calculate the moles of each substance in the reaction and use these to write the balanced equation for the reaction. (A_r Ca = 40, O = 16)

	Ca	O_2	CaO
1 Calculate the number of moles of each substance	$\frac{mass}{M_r} = \frac{20}{40} = 0.50$ mol	$\frac{mass}{M_r} = \frac{8}{32} = 0.25$ mol	$\frac{mass}{M_r} = \frac{28}{56} = 0.50$ mol
2 Divide each number of moles by the smallest one to give the simplest whole number ratio	$\frac{0.50}{0.25} = 2$	$\frac{0.25}{0.25} = 1$	$\frac{0.50}{0.25} = 2$

therefore, the balanced equation is: **2Ca + O_2 → 2CaO**

- You might be given the masses of the reactants and asked to work out the molar ratio in which they react. This is calculated in the same way.

Example

2.70 g of aluminium (Al) reacts with 10.65 g of chlorine (Cl_2). Calculate the moles of aluminium and chlorine and use these to find the simplest molar ratio in which they react. (A_r Al = 27, Cl = 35.5)

	Al	Cl_2
1 Calculate the number of moles of each substance	$\frac{mass}{M_r} = \frac{2.7}{27} = 0.10$ mol	$\frac{mass}{M_r} = \frac{10.65}{71} = 0.15$ mol
2 Divide each number of moles by the smallest one to give the simplest whole number ratio	$\frac{0.10}{0.10} = 1$	$\frac{0.15}{0.10} = 1.5$
3 If this does not give a whole number ratio, you may need to times by a number such as 2 or 3 until you get a whole number ratio	$1 \times 2 = 2$	$1.5 \times 2 = 3$

therefore, they react in the ratio: **2Al + 3Cl_2**

Now test yourself

TESTED ☐

9 14.0 g of nitrogen (N_2) reacts with 3.0 g of hydrogen (H_2) to form 17.0 g of ammonia (NH_3). Calculate the moles of each substance in the reaction and use these to write the balanced equation for the reaction. (A_r N = 14, H = 1)

10 31.2 g of potassium (K) reacts with 6.4 g of oxygen (O_2) to make potassium oxide. Calculate the moles of potassium and oxygen and use these to find the simplest molar ratio in which they react. (A_r K = 39, O = 16)

Answers online

Limiting reactants

- Chemical reactions stop when the reactants have been used up. Sometimes, one of the reactants is used up before the other one, thus stopping the reaction and leaving some of the other reactant unused.
- The reactant that runs out first is called the **limiting reactant**. The other reactant is said to be in excess as some is left over.
- It is common to use an excess of one of the reactants to ensure that the other reactant is all used up.

> **Limiting reactant** is the reactant that is used up first.

Example

What mass of calcium sulfide is formed when 20.0 g of calcium reacts with 19.2 g of sulfur? (A_r Ca = 40, S = 32)

$$Ca + S \rightarrow CaS$$

$$\text{moles of Ca} = \frac{\text{mass}}{M_r} = \frac{20}{40} = 0.50 \text{ mol}$$

$$\text{moles of S} = \frac{\text{mass}}{M_r} = \frac{19.2}{32} = 0.60 \text{ mol}$$

therefore, Ca is the limiting reagent and so 0.50 mol of Ca reacts with 0.50 mol of S to make 0.50 mol of CaS.

mass of CaS = M_r × moles = 72 × 0.50 = 36 g

> **Typical mistake**
>
> You must calculate the number of moles and consider the balanced equation to see which is the limiting reactant. You cannot just look at the mass of each reactant.

Now test yourself

TESTED ☐

11 When 16 g of magnesium reacts with 16 g of bromine, calculate which is the limiting reactant and the mass of magnesium bromide formed. (A_r Mg = 24, Br = 80)

$$Mg + Br_2 \rightarrow MgBr_2$$

12 When 3.6 g of aluminium reacts with 4.0 g of iron oxide, calculate which is the limiting reactant and the mass of iron formed. (A_r Al = 27, Fe = 56, O = 16)

$$2Al + Fe_2O_3 \rightarrow Al_2O_3 + 2Fe$$

Answers online

Concentration of solutions

● The concentration of a solution can be measured in grams per cubic decimetre (g/dm³) or moles per cubic decimetre (mol/dm³).

Table 10.2

units	g/dm³	
equation	(mass triangle: mass over concentration (g/dm³) × volume (dm³))	
	$$\text{concentration (g/dm}^3) = \frac{\text{mass (g)}}{\text{volume (dm}^3)}$$	
example	Find the concentration in g/dm³ of a solution containing 10 g in 2.0 dm³ of solution.	
	$$\text{concentration} = \frac{10}{2.0} = 5 \text{ g/dm}^3$$	

● We often measure volume in cm³ rather than dm³ in the laboratory. There are 1000 cm³ in 1 dm³, and you must convert volumes that are in cm³ to dm³ when using these equations.

$$cm^3 \xrightarrow{\quad \div 1000 \quad} dm^3 \qquad \text{e.g.} \quad 25 \, cm^3 \rightarrow \frac{25}{1000} = 0.025 \, dm^3$$

Now test yourself

TESTED ☐

13 A solution contained 10 g of sodium hydroxide (NaOH) dissolved in 0.5 dm³ of solution. Calculate the concentration in g/dm³.

Answers online

Summary

- Mass is always conserved in chemical reactions.
- The relative formula mass (M_r) is the sum of the mass of all the atoms in a formula.
- **H** One mole of particles contains 6.02×10^{23} particles.
- The mass of one mole of particles of any substance is equal to the M_r in grams.
- The balancing numbers in a chemical equation shows the molar ratio in which the chemicals react.
- In some reactions, an excess of one reactant is used to ensure that all of another reactant is used up.
- The concentration of a solution can be measured in g/dm³.

Exam practice

H 1 How many moles, to 3 significant figures, are there in 1.00 g of sodium oxide (Na_2O)?
(A_r Na = 23, O = 16)

   A 0.016 mol   **B** 0.0161 mol   **C** 0.026 mol   **D** 0.0256 mol   [1 mark]

2 Ammonia (NH_3) is an important chemical used to make fertilisers. Ammonia is made by the reaction of nitrogen from the air with hydrogen. (A_r H = 1, N = 14)

$$N_2 + 3H_2 \rightarrow 2NH_3$$

   (a) What is the maximum mass of ammonia that can be formed from 60 g of hydrogen? [3 marks]
   (b) In an experiment, 100 g of ammonia was formed from 60 g of hydrogen. Calculate the percentage yield. [2 marks]
   (c) Give two reasons why the percentage yield is less than 100%. [2 marks]

3 2.3 g of sodium (Na) reacts with 1.9 g of fluorine (F_2) to make sodium fluoride (NaF) and no other product.
   (a) Using the law of conservation of mass, what mass of sodium fluoride is made in this reaction? [1 mark]
   H (b) Calculate the moles of each substance in the reaction and use these to write the balanced equation for the reaction. (A_r Na = 23, F = 19) [4 marks]

4 In a fuel cell, hydrogen reacts with oxygen to release electrical energy.

$$2H_2(g) + O_2(g) \rightarrow 2H_2O(l)$$

   (a) What volume of oxygen reacts with 12 dm³ of hydrogen, with both gases at room temperature and pressure? [1 mark]
   (b) Calculate the mass of the hydrogen gas used in this reaction. (A_r H = 1) [2 marks]

5 Calcium reacts with chlorine to form calcium chloride. When 10.0 g of calcium (Ca) reacts with 10.0 g of chlorine (Cl_2), calculate which is the limiting reactant and the mass of calcium chloride formed. (A_r Ca = 40, Cl = 35.5) [5 marks]

$$Ca + Cl_2 \rightarrow CaCl_2$$

6 White vinegar is a solution that contains ethanoic acid (CH_3COOH). The concentration of the ethanoic acid in the vinegar was found by titration. In each titration, 25.0 cm³ of the white vinegar was titrated against 0.100 mol/dm³ sodium hydroxide (NaOH) solution. The results are shown in the table.
(A_r H = 1, C = 12, O = 16)

$$CH_3COOH + NaOH \rightarrow CH_3COONa + H_2O$$

titration	1	2	3
volume of sodium hydroxide solution added/cm³	27.6	28.4	27.4

   (a) Describe how the titration should be carried out. [6 marks]
   (b) The student found that the results were repeatable. Explain what this means. [1 mark]
   (c) Calculate the mean volume of sodium hydroxide solution added. Include an estimation of the uncertainty in this mean. [3 marks]
   H (d) Calculate the concentration of the ethanoic acid in mol/dm³ and g/dm³. [5 marks]
   (e) Explain why it would be more difficult to do this titration with brown vinegar. [1 mark]

Answers and quick quiz 10 online

ONLINE

11 Chemical changes

Reactions of metals

The reactivity series of metals

- Metals can be placed in order of reactivity.
- The higher up the series, the more reactive the metal.
- Metals can be put in order by comparing how they react with oxygen, water and acids.

Metal		Reaction with oxygen	Reaction with water	Reaction with acids	
Potassium	K	Burns to form oxide	Reacts and gives off H_2(g)	Reacts violently and gives off H_2(g)	High reactivity
Sodium	Na				
Lithium	Li				
Calcium	Ca			Reacts and gives off H_2(g)	
Magnesium	Mg				
Aluminium	Al				
Carbon	C				
Zinc	Zn	Reacts to form oxide when heated (metal powder burns)	No reaction	Reacts slowly and gives off H_2(g)	Medium reactivity
Iron	Fe				
Tin	Sn				
Lead	Pb				
Hydrogen	H				
Copper	Cu	Forms oxide when heated		No reaction	Low reactivity
Silver	Ag	No reaction			
Gold	Au				
Platinum	Pt				

Most reactive ↑ Least reactive

Figure 11.1

Exam tip

You do not need to learn all these details, but you should know: whether a metal is high, medium or low reactivity; and how high, medium and low metals tend to react with oxygen, water and acids.

Reaction with oxygen

- Most metals react with oxygen.

metal + oxygen → metal oxide

e.g. magnesium + oxygen → magnesium oxide

$$2Mg + O_2 → 2MgO$$

- High-reactivity metals burn in oxygen.
- Mid-reactivity metals react but do not burn. A layer of the metal oxide forms on the surface of the metal.

Reaction with water

- High-reactivity metals react with cold water.

metal + water → metal hydroxide + hydrogen

e.g. sodium + water → sodium hydroxide + hydrogen

$$2Na + 2H_2O \rightarrow 2NaOH + H_2$$

- Potassium, sodium and lithium are in Group 1 of the periodic table (see Chapter 8). They float on the water when they react. Potassium catches fire with a lilac flame. Sodium sometimes catches fire with a yellow-orange flame.

Reaction with acids

- Metals that are more reactive than hydrogen react with acids to form a salt and hydrogen.

metal + acid → salt + hydrogen

e.g. magnesium + sulfuric acid → magnesium sulfate + hydrogen

$$Mg + H_2SO_4 \rightarrow MgSO_4 + H_2$$

Displacement reactions

- A more reactive metal can displace a less reactive metal from a compound.
- For example, copper is more reactive than silver and so will displace silver from a solution of the compound silver nitrate.

> **Displacement reaction**
> a reaction where a more reactive element takes the place of a less reactive element in a compound

word equation: copper + silver nitrate → copper nitrate + silver

balanced equation: $Cu(s) + 2AgNO_3(aq) \rightarrow Cu(NO_3)_2(aq) + 2Ag(s)$

H ● **showing ions separately: $Cu + 2Ag^+ + 2NO_3^- \rightarrow Cu^{2+} + 2NO_3^- + 2Ag$**

final ionic equation: $Cu + 2Ag^+ \rightarrow Cu^{2+} + 2Ag$

What happens when metals react

- When metals react, they are giving away electrons in order to achieve a stable noble gas electron arrangement. This means that the metal atoms become positive ions when they react.
- The more easily a metal gives away its electrons, the more reactive it is.

Oxidation and reduction

Table 11.1

	oxidation	reduction
in terms of oxygen	gain of oxygen	loss of oxygen
in terms of electrons	loss of electrons	gain of electrons

(H) ● When metals react, they are oxidised as they are losing electrons.
● For example, when Mg atoms react with acids, they lose 2 electrons each to form Mg^{2+} ions. This can be written as a half equation:
$Mg \rightarrow Mg^{2+} + 2e^-$ (which can also be written as $Mg - 2e^- \rightarrow Mg^{2+}$)

> **Exam tip**
>
> Remember OIL RIG to work out if a substance is oxidised or reduced in terms of electrons.

● In a displacement reaction, the metal is oxidised but the metal ions in the compound are reduced.
● Reactions in which one substance is oxidised and another one is reduced are called **redox** reactions.

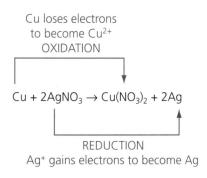

Cu loses electrons
to become Cu^{2+}
OXIDATION

$Cu + 2AgNO_3 \rightarrow Cu(NO_3)_2 + 2Ag$

REDUCTION
Ag^+ gains electrons to become Ag

Figure 11.3

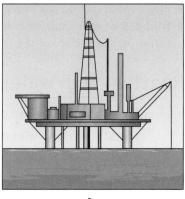

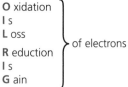

O xidation
I s
L oss
R eduction
I s
G ain
$\Big\}$ of electrons

Figure 11.2

> **Redox reaction:** a reaction in which both reduction and oxidation take place

Metal extraction

 REVISED

● A few very unreactive metals, such as gold, are found as elements on Earth.
● Most metals are found in compounds and have to be extracted by chemical reactions. Compounds from which a metal can be extracted for profit are called **ores**. Most ores are metal oxides.
● Metals that are less reactive than carbon (e.g. iron) are extracted by heating the ore with carbon in a displacement reaction.
● Metals that are more reactive than carbon (e.g. aluminium) have to be extracted by electrolysis which is expensive due to the high cost of the electricity.
● When a metal is produced from a metal oxide ore, reduction takes place as the oxygen is removed.
● However, all extraction reactions are reduction as the metal ions are gaining electrons to form the metal.

	Metal		Method of extraction
Most reactive	Potassium	K	Electrolysis
	Sodium	Na	
	Lithium	Li	
	Calcium	Ca	
	Magnesium	Mg	
	Aluminium	Al	
	Carbon	C	
	Zinc	Zn	Heat with carbon
	Iron	Fe	
	Tin	Sn	
	Lead	Pb	
	Copper	Cu	
	Silver	Ag	Metals found as elements
	Gold	Au	
Least reactive	Platinum	Pt	

Figure 11.4

Now test yourself

1 In each of the following reactions, describe what you would see and write a word equation.
   (a) potassium + water
   (b) magnesium + oxygen
   (c) zinc + sulfuric acid
2 Calcium and zinc both react with hydrochloric acid, but calcium is more reactive.
   (a) Describe what happens, in terms of electrons, to the calcium and zinc atoms when they react with hydrochloric acid.
   (b) Explain, in terms of electrons, why calcium is more reactive than zinc.
(H) 3 When an iron nail is placed in a solution of copper sulfate, the nail becomes coated with a layer of copper due to a displacement reaction.

   Fe(s) + CuSO$_4$(aq) → Cu(s) + FeSO$_4$(aq)

   (a) Is the iron oxidised or reduced in this reaction? Explain your answer.
   (b) Write an ionic equation for this reaction.
4 Aluminium metal is extracted by electrolysis from the ore bauxite, which contains aluminium oxide.
   (a) Why can aluminium not be extracted by heating the aluminium oxide with carbon?
(H) (b) Give two reasons why this extraction is a reduction reaction.

Answers online

Acids and alkalis

What are acids and alkalis?

- An acid is a substance that produces hydrogen ions, H$^+$(aq), when added to water.
- An alkali is a substance that produces hydroxide ions, OH$^-$(aq), when added to water.

The pH scale

- The pH scale is a measure of the acidity or alkalinity of a solution. The pH depends on the concentration of H$^+$(aq) ions.
- A neutral solution has a pH of 7. Acids have a pH of less than 7. Alkalis have a pH greater than 7.
- The pH of a solution can be measured using universal indicator or with a pH probe.

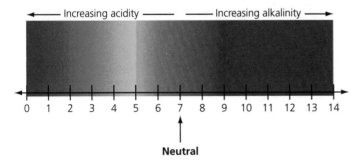

Figure 11.5 **The pH scale and the colours for universal indicator**

(H) ● As the pH decreases by one unit, the concentration of H$^+$(aq) ions increases by a factor of 10. For example, a solution with pH 1 will have a concentration of H$^+$(aq) ions that is 100 times greater than a solution with pH 3.

ⒽStrong and weak acids

- A strong acid is one in which all the molecules break down into ions when added to water (e.g. sulfuric acid, hydrochloric acid, nitric acid). For example, when HCl molecules are added to water they all react to form $H^+(aq)$ ions:

$$H — Cl\,(g) \xrightarrow{\;H_2O\;} H^+(aq) \;+\; Cl^-(aq)$$

Figure 11.6

- A weak acid is an acid in which only a small fraction of the molecules break down into ions (e.g. ethanoic acid, citric acid, carbonic acid).
- The terms strong and weak refer to the fraction of the molecules that break up into ions when added to water. This should not be confused with the terms dilute and concentrated which refer to the amount of acid dissolved in the solution.

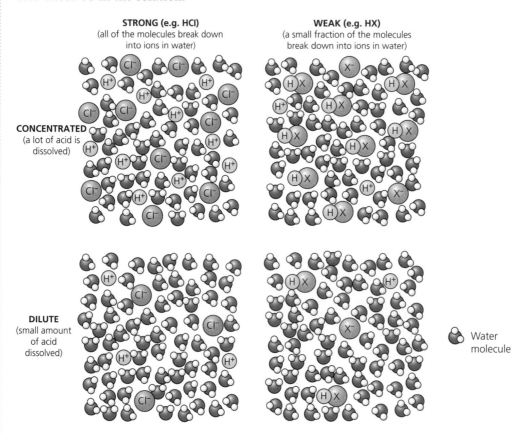

Figure 11.7

Bases

- Bases are substances that react with acids to form a salt and water (and sometimes carbon dioxide also).
- Common bases are:
  - ○ metal oxides (e.g. copper oxide)
  - ○ metal hydroxides (e.g. sodium hydroxide)
  - ○ metal carbonates (e.g. calcium carbonate).
- Alkalis are bases that dissolve in water to form hydroxide ions, $OH^-(aq)$.
- Some metals react with acids to form a salt, but they are not bases as they form a salt and hydrogen (not water).

Salts

- Salts are ionic compounds made when acids react.
- The positive ion in the salt comes from the substance the acid reacts with. For example, a salt would get:
  - Cu^{2+} ions from the base copper oxide (CuO);
  - Na^+ ions from the base sodium hydroxide (NaOH);
  - Mg^{2+} ions from the metal magnesium (Mg).
- The negative ion in the salt comes from the acid. For example, a salt would get:
  - Cl^- ions from hydrochloric acid (HCl), making a chloride salt;
  - SO_4^{2-} ions from sulfuric acid (H_2SO_4), making a sulfate salt;
  - NO_3^- ions from nitric acid (HNO_3), making a nitrate salt.

Reactions of acids

Table 11.2

metals	general	METAL + ACID → SALT + HYDROGEN
	example	zinc + sulfuric acid → zinc sulfate + hydrogen $Zn + H_2SO_4 \rightarrow ZnSO_4 + H_2$
	notes	Acids react with metals that are more reactive than hydrogen to form a salt and hydrogen. When acids react with these metals (e.g. magnesium, zinc and iron), there is fizzing as hydrogen gas is formed. These are redox reactions (electron transfer) as the metal is oxidised to form metal ions and the hydrogen ions are reduced to form hydrogen gas.

metal hydroxides	general	METAL HYDROXIDE + ACID → SALT + WATER
	example	sodium hydroxide + nitric acid → sodium nitrate + water $NaOH + HNO_3 \rightarrow NaNO_3 + H_2O$
	notes	When acids react with metal hydroxides, the ionic equation is always: $H^+(aq) + OH^-(aq) \rightarrow H_2O(l)$ This is a neutralisation reaction (proton transfer) as the $H^+(aq)$ ions are used up.

metal oxides	general	METAL OXIDE + ACID → SALT + WATER
	example	copper oxide + sulfuric acid → copper sulfate + water $CuO + H_2SO_4 \rightarrow CuSO_4 + H_2O$
	notes	The metal oxides are often insoluble in water and so the reaction mixture needs warming. This is a neutralisation reaction (proton transfer) as the $H^+(aq)$ ions are used up.

metal carbonates	general	METAL CARBONATE + ACID → SALT + WATER + CARBON DIOXIDE
	example	calcium carbonate + hydrochloric acid → calcium chloride + water + carbon dioxide $CaCO_3 + 2HCl \rightarrow CaCl_2 + H_2O + CO_2$
	notes	When acids react with carbonates, there is fizzing as carbon dioxide gas is formed. This is a neutralisation reaction (proton transfer) as the $H^+(aq)$ ions are used up.

Making soluble salts

- Salts are important substances that have many uses. For example, many medicines are salts.
- We need to be able to make pure samples of salts.

Stage 1 – THE REACTION

- react the acid with an insoluble substance (e.g. a metal, metal oxide, metal carbonate or metal hydroxide) to produce the desired salt
- the mixture may need to be heated
- add the insoluble substance until it no longer reacts

Stage 2 – FILTER OFF THE EXCESS

- filter off the left-over metal / metal oxide / metal carbonate / metal hydroxide

Stage 3 – CRYSTALLISE THE SALT

- heat the solution to evaporate some water until crystals start to form
- leave the solution to cool down – more crystals will form as it cools
- filter off and wash the crystals of the salt
- allow the crystals to dry

Figure 11.8

5 Name the salt formed in the following reactions.
   (a) zinc + hydrochloric acid
   (b) nickel oxide + nitric acid
   (c) potassium carbonate + sulfuric acid
6 Give the formula of the ion that makes solutions:
   (a) acidic
   (b) alkaline
7 Work out the formula of the following salts. (The formulae of some ions have been given: sulfate = SO_4^{2-}, nitrate = NO_3^-)
   (a) sodium sulfate
   (b) magnesium nitrate
   (c) lithium chloride
8 Barium hydroxide is an alkali that will react with hydrochloric acid.
   (a) Name the type of reaction that takes place.
   (b) Balance the equation for the reaction:

$$Ba(OH)_2(aq) + HCl(aq) \rightarrow BaCl_2(aq) + H_2O(l)$$

   (c) What does the symbol (aq) mean?
   (d) Write an ionic equation for this reaction.
H 9 A large volume of water was added to 10 cm³ of an acid of pH 4. The new solution has a volume of 1000 cm³. What is the pH of this new solution?

Answers and quick quizzes at **www.hoddereducation.co.uk/myrevisionnotesdownloads**

10 Iron tablets are given to patients who have an iron deficiency. The tablets contain the salt iron(II) sulfate which can be made by reacting iron with sulfuric acid. Iron is added to sulfuric acid until all the acid has reacted.
    (a) How could you tell the acid had all reacted?
    (b) How is the left-over iron removed?
    (c) The formula of iron(II) sulfate is $FeSO_4$. Write a balanced equation for the reaction.
    (d) Explain why this reaction is a redox reaction.

Answers online

Required practical 8

Making a soluble salt

Soluble salts can be made in the laboratory by reacting an acid with an insoluble substance, such as a suitable metal, metal oxide, metal carbonate or metal hydroxide.

AIM: To make a pure sample of the salt copper(II) sulfate by reaction of copper(II) oxide with sulfuric acid.

Table 11.3

instructions	comments
1 Safety spectacles should be worn throughout the experiment.	● The sulfuric acid is an irritant/corrosive. It could splash or spit into eyes during this experiment and cause damage.
2 Use a measuring cylinder to place 50 cm³ of sulfuric acid in a beaker.	● A measuring cylinder, rather than a pipette or burette, is suitable here as the exact volume is not important.
3 Heat the beaker using a Bunsen burner. Do not allow the acid to boil.	● The reaction needs to be warmed to take place. However, it would be dangerous to have hot boiling acid that could spit out of the beaker.
4 Add copper(II) oxide using a spatula and stir. The black powder will react and form a blue solution of copper(II) sulfate. After each spatula of powder has reacted, add another one. Keep doing this until there is left-over black copper(II) oxide powder.	● We keep adding copper(II) oxide until all the acid has been used up. This ensures that there will be no left-over acid mixed in with the copper(II) sulfate.
5 Filter the mixture to remove the left-over copper(II) oxide.	● The filtrate is copper(II) sulfate solution. The residue left on the filter paper is the unreacted copper(II) oxide. This ensures that there is no copper(II) oxide mixed in with the copper sulfate solution.
6 The copper(II) sulfate solution is placed in an evaporating basin. It is heated on top of a water bath so that some of the water evaporates. It is heated until crystals of copper(II) sulfate are seen around the edges of the basin.	● A water bath provides heat in a more gentle way than heating directly with a Bunsen burner. ● An alternative is to use an electric heater in which water is heated by a heating element. This would be useful for heating a flammable substance.
7 The basin is left to cool. As it cools, crystals of copper sulfate form.	● Salts are less soluble in water at lower temperature. As the copper sulfate solution cools, it cannot all remain dissolved and so some crystallises out. ● The water is not all boiled off because salts often have water in the crystals.
8 The crystals are filtered off. They are then left to dry.	

Table 11.3 *continued*

Glass rod Beaker Unreacted copper(II) oxide Tripod HEAT Heat-proof mat See instruction 4	Filter funnel Filter paper Residue (unreacted copper(II) oxide) Conical flask Filtrate See instruction 5	Evaporating basin Copper sulfate solution Boiling water Beaker (water bath) HEAT Heat-proof mat See instruction 6

Electrolysis

What is electrolysis?

- **Electrolysis** is the decomposition of ionic compounds using electricity.
- Ionic compounds are solids at room temperature and so cannot conduct electricity.
- However, when molten or dissolved they can conduct and so electrolysis can be done.
- Two electrodes, connected to a supply of electricity, are placed in the molten or dissolved compound.
- The molten or dissolved compound is called the electrolyte.

> **Electrolysis** the decomposition of an ionic compound using electricity

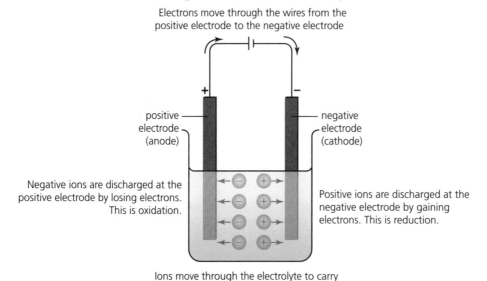

Electrons move through the wires from the positive electrode to the negative electrode

+ positive electrode (anode)

− negative electrode (cathode)

Negative ions are discharged at the positive electrode by losing electrons. This is oxidation.

Positive ions are discharged at the negative electrode by gaining electrons. This is reduction.

Ions move through the electrolyte to carry electric charge. Ions are attracted to the electrode of opposite charge (e.g. positive ions are attracted to the negative electrode)

Figure 11.9

Electrolysis of molten ionic compounds

- Electrolysis of a simple molten ionic compound separates the metal and non-metal that were bonded together in the compound.

Answers and quick quizzes at **www.hoddereducation.co.uk/myrevisionnotesdownloads**

Half equations can be written to show the process at each electrode. For example:

Table 11.4

compound	molten lead bromide ($PbBr_2$)	molten aluminium oxide (Al_2O_3)
ions present	Pb^{2+}, Br^-	Al^{3+}, O^{2-}
– electrode (cathode)	$Pb^{2+} + 2e^- \rightarrow Pb$ (reduction)	$Al^{3+} + 3e^- \rightarrow Al$ (reduction)
+ electrode (anode)	$2Br^- - 2e^- \rightarrow Br_2$ (oxidation)	$2O^{2-} - 4e^- \rightarrow O_2$ (oxidation)

Metal extraction

- Reactive metals, such as aluminium, are extracted by the electrolysis of molten compounds.
- Metals extracted by electrolysis are expensive due to the cost of the electricity and the heat energy to melt the compound.
- In the extraction of aluminium, the aluminium oxide is mixed with cryolite as this lowers the melting point (mixture melts at 950°C, aluminium oxide melts at over 2000°C). This reduces the cost of heat energy.
- The aluminium is formed at the negative electrode as a liquid due to the high temperature.
- The oxygen formed reacts with the positive electrode which is made of graphite. Due to this, the positive electrode needs to be replaced regularly.

Exam tip

Half equations in which electrons are lost can be written in two ways, either with the electrons being taken away from the left or shown with the products on the right. For example:

$$2O^{2-} - 4e^- \rightarrow O_2$$

or

$$2O^{2-} \rightarrow O_2 + 4e^-$$

Typical mistake

Many students confuse whether to add or take electrons to/from ions to discharge them. Remember that electrons are negative. To discharge a positive ion, electrons must be added. To discharge a negative ion, electrons must be taken away.

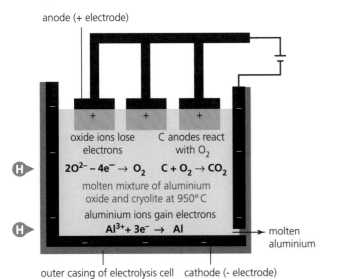

anode (+ electrode)

oxide ions lose electrons
$$2O^{2-} - 4e^- \rightarrow O_2$$

C anodes react with O_2
$$C + O_2 \rightarrow CO_2$$

molten mixture of aluminium oxide and cryolite at 950°C

aluminium ions gain electrons
$$Al^{3+} + 3e^- \rightarrow Al$$

molten aluminium

outer casing of electrolysis cell cathode (- electrode)

Figure 11.10

Electrolysis of aqueous ionic compounds

REVISED

- When an ionic compound is dissolved in water, there are also some H^+(aq) and OH^-(aq) ions present due to the breakdown of water molecules. These ions can be discharged instead of those from the ionic compound.
- At each electrode, the ion that is easier to discharge is the one discharged.
- At the negative electrode, either H^+(aq) ions from the water or metal ions from the compound are discharged. Ions of low-reactivity metals (metals that are less reactive than hydrogen) are discharged. Ions of metals that are more reactive than hydrogen are not discharged, with the H^+(aq) ions discharged instead.

- At the positive electrode, either $OH^-(aq)$ ions from the water or negative ions from the compound are discharged. $Cl^-(aq)$, $Br^-(aq)$ and $I^-(aq)$ ions are discharged rather than $OH^-(aq)$. However, for most other negative ions, the $OH^-(aq)$ ions are discharged.
- $H_2(g)$ is formed if $H^+(aq)$ ions are discharged. $O_2(g)$ is formed if $OH^-(aq)$ ions are discharged.

Table 11.5

electrode	– electrode (cathode)	+ electrode (anode)
ions discharged	+ ions discharged	– ions discharged
sodium chloride, NaCl(aq)	$H^+(aq)$ discharged not $Na^+(aq)$ $\quad$ (H) $2H^+ + 2e^- \rightarrow H_2$ (reduction)	$Cl^-(aq)$ discharged not $OH^-(aq)$ $\quad$ (H) $2Cl^- - 2e^- \rightarrow Cl_2$ (oxidation)
silver nitrate, AgNO₃(aq)	$Ag^+(aq)$ discharged not $H^+(aq)$ $\quad$ (H) $Ag^+ + e^- \rightarrow Ag$ (reduction)	$OH^-(aq)$ discharged not $NO_3^-(aq)$ $\quad$ (H) $4OH^- - 4e^- \rightarrow O_2 + 2H_2O$ (oxidation)
copper(II) bromide, CuBr₂(aq)	$Cu^{2+}(aq)$ discharged not $H^+(aq)$ $\quad$ (H) $Cu^{2+} + 2e^- \rightarrow Cu$ (reduction)	$Br^-(aq)$ discharged not $OH^-(aq)$ $\quad$ (H) $2Br^- - 2e^- \rightarrow Br_2$ (oxidation)
potassium sulfate, K₂SO₄(aq)	$H^+(aq)$ discharged not $K^+(aq)$ $\quad$ (H) $2H^+ + 2e^- \rightarrow H_2$ (reduction)	$OH^-(aq)$ discharged not $SO_4^{2-}(aq)$ $\quad$ (H) $4OH^- - 4e^- \rightarrow O_2 + 2H_2O$ (oxidation)

Now test yourself

TESTED ☐

11 The electrolysis of molten sodium chloride produces sodium at the cathode and chlorine at the anode. The half equation for the reaction at the cathode is:

$Na^+ + e^- \rightarrow Na$

(a) Why can the electrolysis not be done with solid sodium chloride?
(b) Why do sodium ions move to the cathode.
(c) Are the sodium ions oxidised or reduced? Explain your answer.
(d) Write a half equation for the reaction at the anode.

12 Predict the products at the anode and cathode in the electrolysis of the following substances:
(a) molten calcium bromide
(b) aqueous copper(II) sulfate
(c) aqueous potassium iodide

13 Balance the following half equations:
(a) $Mg^{2+} + e^- \rightarrow Mg$
(b) $I^- - e^- \rightarrow I_2$
(c) $O^{2-} \rightarrow O_2 + e^-$
(d) $Fe^{2+} + e^- \rightarrow Fe$

Answers online

Required practical 9

Electrolysis of aqueous solutions of ionic compounds

AIM: To find out what is made at the electrodes in the electrolysis of aqueous solutions of some ionic compounds.

- The solutions being tested are:
  - copper(II) chloride
  - copper(II) sulfate
  - sodium chloride
  - sodium sulfate

- The possible products from these compounds and how they can be identified are shown in the table:

Table 11.6

electrode	negative electrode (cathode)			positive electrode (anode)	
possible products	copper (Cu)	sodium (Na)	hydrogen (H_2)	chlorine (Cl_2)	oxygen (O_2)
observation	brown coating on the electrode	grey coating on the electrode	bubbles of gas	bubbles of gas that bleach damp litmus paper	bubbles of gas that do not bleach damp litmus paper

- An electrolysis cell is set up using a supply of electricity and graphite electrodes.
- The electricity supply can be from batteries or a power pack.
- The graphite electrodes are used as they conduct electricity but are inert, i.e. they do not react themselves.
- By careful observations and tests, the product at each electrode can be identified.

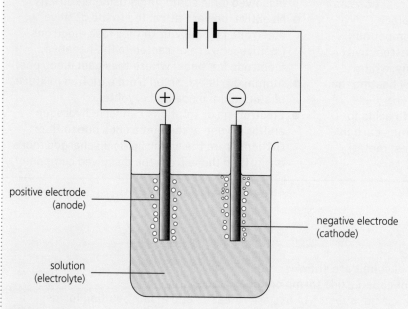

positive electrode (anode)

negative electrode (cathode)

solution (electrolyte)

Figure 11.11

Results

Table 11.7

solution	copper(II) chloride	copper(II) sulfate	sodium chloride	sodium sulfate
negative electrode (cathode)	brown coating on the electrode = copper	brown coating on the electrode = copper	bubbles of gas = hydrogen	bubbles of gas = hydrogen
positive electrode (anode)	bubbles of gas that bleach damp litmus paper = chlorine	bubbles of gas that do not bleach damp litmus paper = oxygen	bubbles of gas that bleach damp litmus paper = chlorine	bubbles of gas that do not bleach damp litmus paper = oxygen

Summary

- Metals can be arranged in order of reactivity. This is known as the reactivity series of metals.
- Most metals react with oxygen, when heated, to form a metal oxide. Reactive metals burn in oxygen.
- Reactive metals react with cold water to form a metal hydroxide and hydrogen.
- Metals that are more reactive than hydrogen react with dilute acids to form a salt and hydrogen.
- A more reactive metal will displace a less reactive metal from its compounds.
- Oxidation is loss of electrons; reduction is gain of electrons (OIL RIG).
- Displacement reactions are redox reactions as both reduction and oxidation take place.
- Most metals are extracted from compounds in ores by chemical reactions. High-reactivity metals are extracted by electrolysis, while most other metals are extracted by heating the ores with carbon.
- Bases are chemicals that react with acids to form a salt and water (and sometimes carbon dioxide). Bases include metal oxides, metal hydroxides and metal carbonates.

- Acids are substances that form $H^+(aq)$ ions when added to water. Their pH is less than 7.
- Alkalis are substances that form $OH^-(aq)$ ions when added to water. Their pH is more than 7. Most alkalis are water soluble metal hydroxides.
- In strong acids, all the molecules break into ions when added to water. In weak acids, only a small fraction of the molecules break into ions when added to water.
- Soluble salts can be made by reacting acids with an excess of metals or insoluble bases. The excess is filtered off and the salt is crystallised from the solution.
- Electrolysis is the decomposition of molten or dissolved ionic compounds using electricity.
- Negative ions are attracted to the positive electrode (anode) where they lose electrons. Positive ions are attracted to the negative electrode (cathode) where they gain electrons.
- Aluminium is extracted from a molten mixture of aluminium oxide and cryolite.
- Electrolysis of solutions can form hydrogen and/or oxygen at the electrodes due to H^+ or OH^- ions from the water being discharged more easily than those from the dissolved compound.

Exam practice

1 Cobalt is a metal. Some reactions of cobalt are shown:
  - when heated in oxygen, a layer of cobalt oxide forms on the surface;
  - when added to warm hydrochloric acid, bubbles of gas are released and a pink solution forms;
  - when a piece of cobalt is added to copper sulfate solution, a displacement reaction takes place and the surface of the cobalt is covered in brown copper metal.
  (a) Write a balanced equation for the reaction of cobalt (Co) with oxygen to form cobalt oxide (CoO). [2 marks]
  (b) Write a word equation for the reaction of cobalt with hydrochloric acid. [2 marks]
  (c) Explain why cobalt displaces copper from copper sulfate solution. [1 mark]

2 Scrap iron can be used to extract copper metal from solutions, such as copper(II) sulfate, in a redox reaction. This reaction takes place because iron is more reactive than copper.

  $CuSO_4(aq) + Fe(s) \rightarrow Cu(s) + FeSO_4(aq)$

  (a) Explain, in terms of electrons, why iron is more reactive than copper. [1 mark]
  (b) Write an ionic equation for this reaction (sulfate ions have the formula SO_4^{2-}). [2 marks]
  (c) Explain why this is a redox reaction. [3 marks]

3 The pH values of some solutions are shown. Which is the most acidic?
  A pH 0   B pH 2   C pH 6   D pH 9   [1 mark]

4 Which one of the following is a weak acid?
  A carbonic acid   B hydrochloric acid   C nitric acid   D sulfuric acid   [1 mark]

5 Hydrochloric acid reacts to form chloride salts. Draw lines to show the correct products for the reactions of magnesium and magnesium oxide with hydrochloric acid. [2 marks]

| magnesium |

| magnesium oxide |

| magnesium chloride + water |

| magnesium chloride + hydrogen |

| magnesium chloride + water + carbon dioxide |

6 Calcium nitrate is a soluble salt used in some fertilisers. It can be made by reaction of calcium carbonate with an acid. A student made some calcium nitrate in the laboratory using the following method.
Step 1 Add solid calcium carbonate to some acid in a beaker until the reaction has finished.
Step 2 Filter.
Step 3 Warm the solution until crystals start to form and then leave to cool.
Step 4 Filter off the crystals of calcium nitrate.
(a) Name the acid that should be used. [1 mark]
(b) Write a word equation for the reaction. [1 mark]
(c) How can you tell when the reaction has finished? [1 mark]
(d) Why was the mixture filtered in step 2? [1 mark]
(e) Nitrate ions have the formula NO_3^-. What is the formula of calcium nitrate? [1 mark]

7 What would be formed at the cathode in the electrolysis of magnesium chloride solution?
A chlorine, B hydrogen, C magnesium, D oxygen [1 mark]

8 Aluminium metal is extracted from the electrolysis of a molten mixture of cryolite and aluminium oxide using graphite electrode. Aluminium is formed at the cathode and oxygen is formed at the anode.
(a) Why must the mixture be molten? [1 mark]
(b) Why is the aluminium oxide mixed with cryolite? [2 marks]
(c) Write a half equation for the formation of aluminium from Al^{3+} ions at the cathode. [1 mark]
(d) Explain why the anode has to be replaced regularly. [1 mark]

9 Chlorine is made by the electrolysis of sodium chloride solution. It is formed at the anode.
(a) Balance the half equation for the formation of chlorine. [1 mark]

$$Cl^- \rightarrow Cl_2 + e^-$$

(b) Explain why the production of chlorine is an oxidation reaction. [1 mark]
(c) Explain why hydrogen rather than sodium is formed at the cathode during this process. [2 marks]

Answers and quick quiz 11 online

ONLINE

12 Energy changes

Exothermic and endothermic reactions

- Energy is conserved in a chemical reaction.
- However, the reactants and products have different amounts of energy and so energy is transferred to/from the surroundings in order to conserve energy.

Table 12.1

energy change	exothermic	endothermic
what happens	thermal energy is transferred from the chemicals to the surroundings	thermal energy is transferred from the surroundings to the chemicals
temperature of surroundings	gets hotter	gets colder
reaction profile	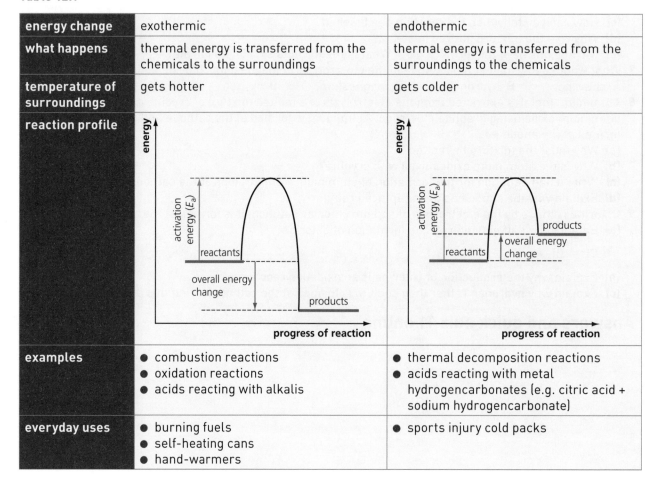	
examples	• combustion reactions • oxidation reactions • acids reacting with alkalis	• thermal decomposition reactions • acids reacting with metal hydrogencarbonates (e.g. citric acid + sodium hydrogencarbonate)
everyday uses	• burning fuels • self-heating cans • hand-warmers	• sports injury cold packs

- Reaction profiles (energy level diagrams) show:
  - the relative energy of the reactants and products
  - the overall energy change
  - the activation energy which is the minimum energy particles need to react when they collide.

Now test yourself

1 Decide whether each of the following reactions is exothermic or endothermic.
  (a) burning petrol in a car engine
  (b) reaction of citric acid and sodium hydrogencarbonate in fizzy sweets
  (c) a reaction in solution that lowers the temperature of the solution from 20°C to 15°C
  (d) neutralisation of sulfuric acid by sodium hydroxide
  (e) a reaction where the reactants have more energy than the products.

Answers online

Required practical 10

Temperature change in reactions

There are many factors that affect the temperature change in a reaction which can be investigated. Reactions can be done in solution in a polystyrene cup (for insulation) and the temperature change measured.

Table 12.2

Practical 10a		
AIM: To see how changing the mass of a reactant affects the temperature change.		
Sodium hydrogencarbonate reacts with hydrochloric acid in an endothermic reaction. $NaHCO_3 + HCl \rightarrow NaCl + H_2O + CO_2$ ● 30 cm³ of 2.0 mol/dm³ HCl was placed in a polystyrene cup and the temperature measured. ● 1 g of solid $NaHCO_3$ was added and the mixture stirred. The drop in temperature was measured. ● The experiment was repeated with 2, 3, 4 and 5 g of $NaHCO_3$.	**mass of NaHCO₃ /g**	**temperature drop /°C**
	1	2
	2	5
	3	6
	4	8
	5	11

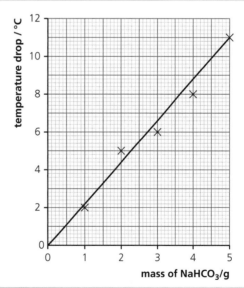

CONCLUSION: The greater the mass of $NaHCO_3$ used, the more the temperature falls.	

Table 12.2 *continued*

Practical 10b

AIM: To see how changing the volume of a reactant affects the temperature change.

Sodium hydroxide reacts with hydrochloric acid in an exothermic reaction.

$$NaOH + HCl \rightarrow NaCl + H_2O$$

- 25 cm³ of 2.0 mol/dm³ NaOH was placed in a polystyrene cup and the temperature measured.
- 40 cm³ of 2.0 mol/dm³ HCl was added 5 cm³ at a time from a burette and the temperature measured.

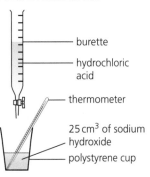

volume of HCl/ cm³	temperature/ °C	temperature rise since the start /°C
0	21	0
5	26	5
10	29	8
15	31	10
20	33	12
25	35	14
30	34	13
35	33	12
40	32	11

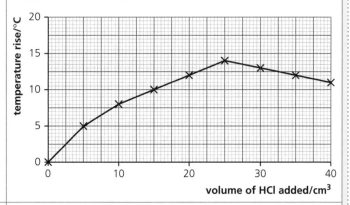

CONCLUSION: The greater the volume of acid added, the greater the temperature rise up until 25 cm³ has been added, but after this the temperature rise reduces.

Some key points about plotting graphs:
- the independent variable goes on the horizontal axis (*x*-axis), with the dependent variable on the vertical axis (*y*-axis)
- the scale on each axis should use more than half the available space
- axes should be labelled including units
- if there is a pattern, then a best fit line should be drawn which could be straight (drawn with a ruler) or curved
- there should be a similar number of points above and below a best fit line
- sometimes there are two best fit lines that meet where the relationship changes.

Types of variables

Table 12.3

variables	description	experiment 4a	experiment 4b
independent	the variable that we change	• mass of NaHCO₃	• volume of acid
dependent	the variable that we measure	• temperature change	• temperature change
control	variables that must be kept the same to make it a fair test	• volume of acid • concentration of acid • particle size of NaHCO₃	• volume of NaOH in cup • concentration of NaOH • concentration of acid

ⓗCalculating energy change in reactions

- Energy is needed to break a chemical bond.
- Energy is released when a chemical bond is made.
- The energy change for a reaction can be calculated using bond energies.

Energy = Energy needed to break − Energy released making
change   bonds in reactants   bonds in products

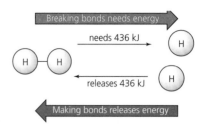

Figure 12.1

Example

Find the energy change in the following reaction using bond energies.

Explain, in terms of bond energies, whether the reaction is exothermic or endothermic.

H—C—H (with H above and H below) + 2 O=O ⟶ O=C=O + 2 H—O—H (water structure)

bond	C–H	O=O	C=O	O–H
bond energy /kJ	412	496	743	463

(a)

bonds broken	energy	bonds made	energy
4 C–H	4(412)	2 C=O	2(743)
2 O=O	2(496)	4 O–H	4(463)
Total	2640 kJ	Total	3338 kJ

energy change = 2640 − 3338 = −698 kJ

(b) the reaction is exothermic as more energy is released making bonds than is needed to break bonds

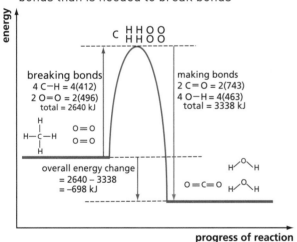

Figure 12.2

Typical mistake

A lot of students refer to energy being needed to make bonds. This is wrong; energy is released when bonds are made.

Now test yourself

2 (a) Calculate the energy change for the reaction of hydrogen with oxygen using the bond energies given. (bond energies: H–H = 436, O=O = 496, O–H = 463 kJ)

2 H—H + O=O ⟶ 2 H—O—H (water structure)

(b) Explain, in terms of bond energies, whether the reaction is exothermic or endothermic.

Answers online

Summary

- There is an energy change in chemical reactions.
- Exothermic reactions transfer thermal energy to the surroundings, which get hotter. Examples include burning fuels, oxidation and acids reacting with alkalis.
- Endothermic reactions transfer thermal energy away from the surroundings, which get colder. Examples include thermal decomposition reactions and acids reacting with metal hydrogencarbonates.
- A reaction profile shows the energy changes during a reaction.
- **H** ● Breaking bonds requires energy. Making bonds releases energy.
- The energy change in a chemical reaction can be found using bond energies.

Exam practice

1 Sodium hydroxide solution was added to hydrochloric acid in a beaker. The temperature rose from 21°C to 27°C.

   (a) Is this reaction endothermic or exothermic? Explain your answer.   [1 mark]

   (b) Sketch the reaction profile for this reaction.   [3 marks]

2 The reaction profile for a reaction is shown.

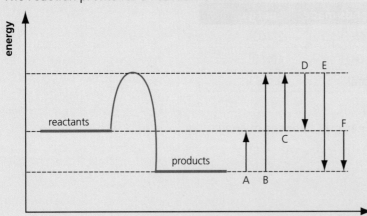

   (a) Give the letter for the overall energy change for this reaction.   [1 mark]

   (b) Give the letter for the activation energy for this reaction.   [1 mark]

   (c) Is this reaction exothermic or endothermic? Explain your answer.   [1 mark]

H 3 **(a)** Calculate the energy change for the reaction of ethene with bromine using the bond energies given. (bond energies: C=C = 612, C–H = 412, Br–Br = 193, C–C = 348, C–Br 276 kJ)   [3 marks]

$$
\underset{\text{H}}{\overset{\text{H}}{|}}\text{C}=\underset{\text{H}}{\overset{\text{H}}{|}}\text{C}\text{—H} + \text{Br—Br} \longrightarrow \text{H—}\underset{\text{Br}}{\overset{\text{H}}{|}}\text{C}\text{—}\underset{\text{Br}}{\overset{\text{H}}{|}}\text{C}\text{—H}
$$

   (b) Explain, in terms of bond energies, whether this reaction is exothermic or endothermic.   [1 mark]

→

4 Copper sulfate solution reacts with zinc powder in an exothermic reaction.
$CuSO_4(aq) + Zn(s) \rightarrow Cu(s) + ZnSO_4(aq)$
An investigation was carried out to see how changing the mass of zinc affects the temperature rise.
In each experiment, zinc powder was added to 40 cm^3 of 0.5 mol/dm^3 copper sulfate solution in a plastic cup.

mass of zinc (g)	0	0.25	0.50	0.75	1.00	1.25	1.50	1.75	2.00
temperature rise (°C)	0	6	10	14	20	24	25	24	25

(a) Identify the independent, dependent and two control variables in this investigation. [4 marks]
(b) Plot a best-fit graph of the temperature change against the mass of zinc. [4 marks]
(c) Describe the relationship shown by these results. [2 marks]
(d) Calculate the number of moles of copper sulfate in each experiment. [1 mark]
(e) Calculate the maximum mass of zinc that can react with this amount of copper sulfate. [2 marks]
(f) Explain why the relationship between the mass of zinc and the temperature rise changes at a point between 1.25 and 1.50 g of zinc. [1 mark]
(g) Write an ionic equation for the reaction taking place. The formulae of the ions involved are: copper ions = Cu^{2+}, zinc ions = Zn^{2+}, sulfate ions = SO_4^{2-} [1 mark]

Answers and quick quiz 12 online

ONLINE

13 The rate and extent of chemical reactions

Rate of reaction

- Some reactions are fast but others are slow. We can measure and control the rate of reactions.

Mean rate of reaction

REVISED

- The rate of a chemical reaction changes during the reaction. Most reactions are fastest at the beginning, slow down and stop.
- The mean rate of reaction can be found by measuring the quantity of reactant used or product formed divided by the time taken.
- The quantity can be:
  - mass in grams, giving rate in g/s
  - volume of gas in cm^3, giving rate in cm^3/s
  - **(H)** amount in moles, giving rate in mol/s

$$\text{mean rate of reaction} = \frac{\text{quantity of reactant used OR product formed}}{\text{time}}$$

> **Example**
>
> In a reaction, 100 cm^3 of gas are formed in 50 s. What is the mean rate of reaction?
>
> $$\text{mean rate of reaction} = \frac{100\ cm^3}{50\ s} = 2\ cm^3/s$$

Reaction rate graphs

REVISED

- Graphs can be drawn to show how the quantity of reactant used or product formed varies with time.

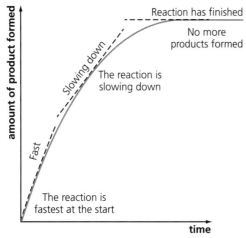

amount of product formed ↑

Reaction has finished

No more products formed

Slowing down

The reaction is slowing down

Fast

The reaction is fastest at the start

time →

Figure 13.1

- The slope (gradient) of the line represents the rate of reaction.
- The steeper the line, the faster the rate of reaction.

Answers and quick quizzes at **www.hoddereducation.co.uk/myrevisionnotesdownloads**

H ● The rate of reaction at any point can be found by drawing a tangent to the line and finding the slope (gradient) of the line.

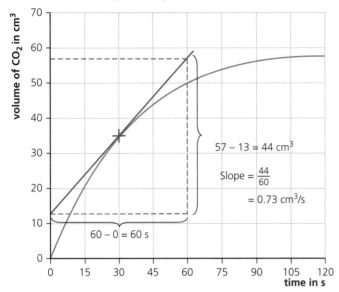

Figure 13.2

Collision theory

● For a chemical reaction to take place, the particles must collide with enough energy to react (this is called a successful collision).
● The **activation energy** is the minimum energy the particles need to react.
● The rate of reaction depends on the frequency of successful collisions.

> **Activation energy** the minimum energy that particles need to react

Why reactions slow down

● Reactions are fastest at the beginning, slow down and eventually stop.
● Reactions slow down over time because there are less reactant particles and so the successful collisions between reactant particles are less frequent. The reaction stops when there are no more particles of one of the reactants.

> **Exam tip**
>
> When you explain why reactions slow down and stop, you must specifically refer to the number of **reactant** particles rather than the number of particles.

The effect of concentration of solutions on rate of reaction

● The higher the concentration of reactants in solution, the faster the rate of reaction.
● This is because particles are closer together and so successful collisions are more frequent.
● If the concentration is doubled, successful collisions are twice as frequent. Therefore, the rate is proportional to the concentration.

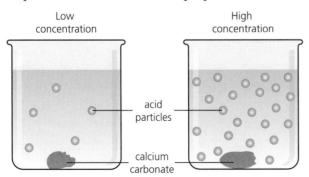

Figure 13.3 Acid reacting with calcium carbonate

> **Typical mistake**
>
> Some students write about the number of collisions. This is wrong – you need to talk about the **frequency** of **successful** collisions.

The effect of pressure of gases on rate of reaction

- The higher the pressure of reactant gases, the faster the rate of reaction.
- This is because particles are closer together and so successful collisions are more frequent.
- If the pressure is doubled, successful collisions are twice as frequent. Therefore, the rate is proportional to the pressure.

Exam tip

Pressure only affects the rate of reactions where one or more of the reactants is a gas. It does not affect reactions where the only gases are products or there are no gases at all.

Low pressure   High pressure

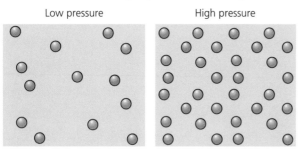

Figure 13.4 Reactions are faster at higher pressure

The effect of surface area of solids on rate of reaction

- The more pieces a solid is broken up into, the greater the surface area.
- For example, if a cube is broken up into smaller cubes, the surface area to volume ratio is greater.

Table 13.1

	one 2 cm × 2 cm × 2 cm cube	eight 1 cm × 1 cm × 1 cm cubes
surface area	area of each face = 2 × 2 = 4 cm² for all 6 faces = 6 × 4 = 24 cm²	area of each face = 1 × 1 = 1 cm² for all 6 faces on each cube = 6 × 1 = 6 cm² for all 8 cubes = 8 × 6 = 48 cm²
volume	volume = 2 × 2 × 2 = 8 cm³	volume of each cube = 1 × 1 × 1 = 1 cm³ for all 8 cubes = 8 × 1 = 8 cm³
ratio	surface area:volume = 24:8 = **3:1**	surface area:volume = 48:8 = **6:1**

- The greater the surface area to volume ratio of solid reactants, the faster the rate of reaction.
- This is because there are more reactant particles on the surface. Therefore, successful collisions are more frequent.
- If the surface area is doubled, successful collisions are twice as frequent. Therefore, the rate is proportional to the surface area.

Large piece of solid – low surface area **Smaller pieces of solid – bigger surface area**

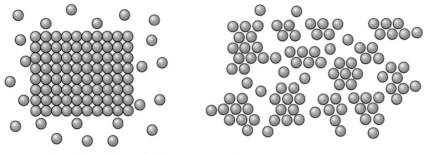

Figure 13.5 **The effect of changing surface area**

The effect of temperature on rate of reaction

- The higher the temperature, the faster the rate of reaction.
- This is because:
  - the particles have more energy and so more of the collisions are successful
  - the particles move faster and so collisions are more frequent.
- If the temperature is doubled, the rate of reaction is much more than doubled as the increasing temperature increases both the success and frequency of collisions. Therefore, rate of reaction is not proportional to temperature.

The effect of catalysts on rate of reaction

- A **catalyst** is a substance that increases the rate of a reaction but is not used up.
- Different reactions have different catalysts.
- Enzymes act as catalysts in biological systems.
- Catalysts do not appear in the overall chemical equation for a reaction as they are not used up.
- Catalysts work by providing a different reaction pathway that has a lower activation energy.

> **Catalyst** a substance that speeds up a reaction without being used up

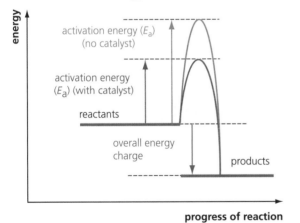

Figure 13.6

Now test yourself

1 Calculate the mean reaction rate in the following reactions.
 (a) 40 cm³ of gas is produced in 10 s
 (b) 0.032 g of gas is given off in 20 s
 (c) 0.0150 mol of a substance is made in 30 s
2 Figure 6.7 shows the volume of hydrogen gas formed against time in three reactions. Which reaction, P, Q or R, is fastest? Explain your answer.
3 The volume of carbon dioxide formed when some calcium carbonate reacts with hydrochloric acid is plotted against time in Figure 13.8.

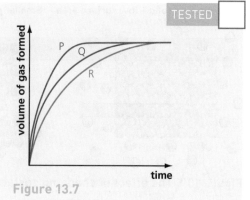

Figure 13.7

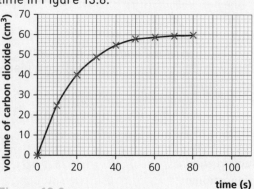

Figure 13.8

 (a) Explain why the reaction slows down and stops.
 (b) Calculate the rate at 20 seconds in cm³/s by drawing a tangent to the line and finding the gradient.
4 Explain each of the following using collision theory.
 (a) 2.0 mol/dm³ sulfuric acid reacts with magnesium faster than 1.0 mol/dm³ sulfuric acid.
 (b) Powdered calcium carbonate reacts with hydrochloric acid faster than chips of calcium carbonate.
 (c) Hydrogen peroxide decomposes faster at 40°C than at 20°C.
 (d) Hydrogen peroxide decomposes faster in the presence of the catalyst manganese(IV) oxide.
5 Calculate the surface area to volume ratio of a cube with 5 cm sides.

Answers online

Required practical 11

The effect of concentration on reaction rate

Table 13.2

Practical 11a	Practical 11b
AIM: To see how changing concentration affects rate by a method using gas volumes.	**AIM:** To see how changing concentration affects rate by a method using turbidity (cloudiness).
Magnesium reacts with hydrochloric acid to form hydrogen gas. The rate can be found by measuring the volume of gas formed in a set time. $$\text{rate} = \frac{\text{volume (cm}^3)}{\text{time (s)}}$$	Sodium thiosulfate solution reacts with hydrochloric acid and forms solid sulfur that is insoluble in the aqueous mixture. The rate can be found by measuring the time taken for the mixture to become so cloudy that you cannot see through it. This can be done by putting the reaction mixture on a piece of paper with a cross. The rate is found as 1000 divided by the time taken (rate is always something divided by time – there is no obvious number to use here so 1000 is used for convenience). $$\text{rate (/s)} = \frac{1000}{\text{time (s)}}$$

Table 13.2 *continued*

Practical 11a	Practical 11b

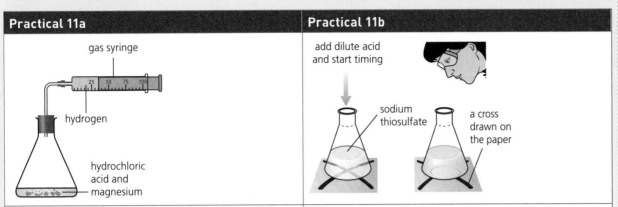

Practical 11a	Practical 11b
• 50 cm³ of 1.0 mol/dm³ HCl was placed in a conical flask attached to a gas syringe. • 5 cm of magnesium ribbon was added. • The volume of gas produced in the first 30 seconds was recorded. • The experiment was repeated with 0.8, 0.6, 0.4 and 0.2 mol/dm³ hydrochloric acid.	• 50 cm³ of 0.25 mol/dm³ sodium thiosulfate solution was placed in a conical flask on a piece of paper with a cross. • 10 cm³ of 2.0 mol/dm³ HCl was added. • The mixture was stirred and the time taken for the mixture to become too cloudy to see the cross measured. • The experiment was repeated with 0.20, 0.15, 0.10 and 0.05 mol/dm³ sodium thiosulfate solution.

concentration of HCl/ mol/dm³	Volume of gas in 30 seconds /cm³	Rate of reaction/ cm³/s
1.0	60	2.00
0.8	45	1.50
0.6	35	1.17
0.4	22	0.73
0.2	10	0.33

concentration of sodium thiosulfate/ mol/dm³	time to become too cloudy to see cross/s	rate of reaction/s
0.25	58	17.2
0.20	72	13.9
0.15	98	10.2
0.10	151	6.6
0.05	320	3.1

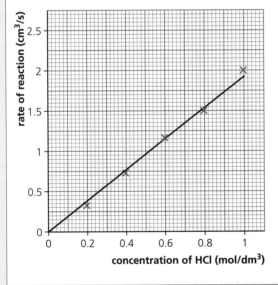

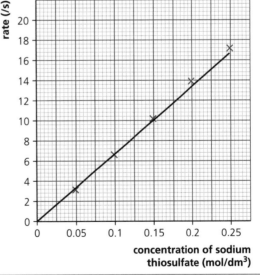

Proportional relationships

- In these experiments, the rate of reaction is proportional to the concentration.
- In each case, this means that doubling the concentration will double the rate; for example, making the concentration three times bigger will make the rate three times faster.
- Where there is a proportional relationship, the graph of the two variables will be a straight-line graph that goes through where the axes meet at (0,0).

Reversible reactions and dynamic equilibrium

Reversible reactions

- Some reactions are reversible. This means that once the products have been made they can turn back into the reactants.

$$\text{reactants} \xrightleftharpoons[\text{reverse reaction}]{\text{forward reaction}} \text{products}$$

e.g. hydrated copper sulfate $\rightleftharpoons$ **anhydrous copper sulfate + water**
(blue)   (white)

e.g. ammonium chloride $\rightleftharpoons$ **ammonia + hydrogen chloride**

- If the forward reaction is exothermic, the reverse reaction is endothermic.
- If the forward reaction is endothermic, the reverse reaction is exothermic.
- For example, if the forward reaction has an energy change of $-76\,\text{kJ}$, then the reverse reaction would have an energy change of $+76\,\text{kJ}$.

> **Exam tip**
>
> If you need to know whether a reaction is exothermic or endothermic you will probably be told unless you can work it out. You are usually told about the energy change for the forward reaction.

Dynamic equilibrium

- In a closed system, where no chemicals can get in or out, a reversible reaction can reach a state of dynamic equilibrium.
- At equilibrium, both the forward and reverse reactions are taking place simultaneously but at exactly the same rate.
- A good analogy of a system in a state of dynamic equilibrium is somebody walking up an escalator that is moving down. It is a dynamic equilibrium when the person is walking up at exactly the same rate as the escalator is moving down.

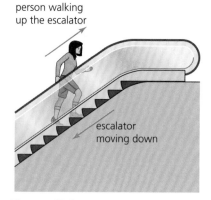

person walking up the escalator

escalator moving down

Figure 13.9

Le Châtelier's Principle

The position of an equilibrium

- If the position of an equilibrium lies to the left, it means that there are more of the reactants than products present at equilibrium.
- If the position of an equilibrium lies to the right, it means that there are more of the products than reactants present at equilibrium.

Changing the position of an equilibrium

- If a change is made to the conditions of a system at equilibrium, then the position of the equilibrium moves to oppose that change in conditions.

H Table 13.3

factor changed	increase	decrease	notes
temperature	• if the temperature is increased, the equilibrium position moves in the direction of the endothermic reaction to reduce the temperature	• if the temperature is decreased, the equilibrium position moves in the direction of the exothermic reaction to increase the temperature	• exothermic reactions increase the temperature • endothermic reactions decrease the temperature
pressure	• if the pressure is increased, the equilibrium position moves towards the side with fewer gas molecules to reduce the pressure	• if the pressure is decreased, the equilibrium position moves towards the side with more gas molecules to increase the pressure	• the more gas molecules, the greater the pressure
concentration	• if the concentration of a substance is increased, the position of the equilibrium moves to reduce it by doing more of the reaction that uses it up	• if the concentration of a substance is decreased, the position of the equilibrium moves to increase it by doing more of the reaction that makes it	

Example

Hydrogen can be made by reaction of methane with steam.

forward reaction is endothermic →

| methane | + | steam | ⇌ | hydrogen | + | carbon monoxide |
| $CH_4(g)$ | + | $H_2O(g)$ | ⇌ | $3H_2(g)$ | + | $CO(g)$ |

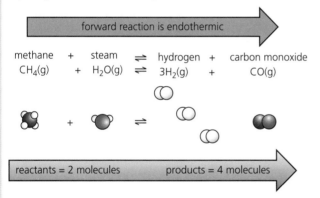

reactants = 2 molecules   products = 4 molecules →

Figure 13.10

Table 13.4

	what happens	effect on yield of hydrogen
increase temperature	equilibrium position moves right in endothermic direction to lower the temperature	more hydrogen formed
increase pressure	equilibrium position moves left to side with less gas molecules to lower the pressure	less hydrogen formed
increase concentration of steam	equilibrium position moves right to remove added steam	more hydrogen formed

Exam tip

In an equilibrium with the same number of gas molecules on each side, the position of the equilibrium will not move if the pressure is changed

Now test yourself

6 Ethene (C_2H_4) reacts with steam to make ethanol (C_2H_5OH) in a reaction that reaches a state of equilibrium. The forward reaction is exothermic.

$$C_2H_4(g) + H_2O(g) \rightleftharpoons C_2H_5OH(g)$$

(a) What would happen to the yield of ethanol if the reaction temperature was increased? Explain your answer.

(b) What would happen to the yield of ethanol if the pressure was increased? Explain your answer.

(c) What would happen to yield of ethanol if more steam was added to the mixture? Explain your answer.

Answers online

Summary

- For a chemical reaction to take place, particles must collide with enough energy to react (this is called a successful collision).
- The activation energy is the minimum amount of energy that particles need to react.
- The rate of reaction depends on the frequency of successful collisions.
- Reaction rate is increased by:
  - higher concentration of a reactant in solution – this is because the particles are closer together and so successful collisions are more frequent;
  - higher pressure of reactant gases – this is because particles are closer together and so successful collisions are more frequent;
  - greater surface area of a solid reactant – this is because more particles are on the surface and so successful collisions are more frequent;
  - higher temperature – this is because particles have more energy and move faster and so more of the collisions are successful and collisions are more frequent;

- use of a catalyst (a substance that increases the rate of reaction without being used up) – catalysts work by providing an alternative path for the reaction with a lower activation energy
- The steepness (gradient) of a graph of mass/volume/moles of product against time gives the rate of reaction. The steeper the line, the faster the reaction.
- Reactions are fastest at the start and then slow down until they stop. They slow down as there are less reactant particles and so successful collisions between reactant particles are less frequent.
- Many chemical reactions are reversible. This means that the reactants can be reformed from the products.
- In a closed system (where nothing can get in or out), a reversible reaction can reach a state of dynamic equilibrium where both the forwards and reverse reactions are taking place simultaneously and at the same rate.
- (H) If the conditions of a system at equilibrium are changed, the position of the equilibrium moves to oppose that change.

Exam practice

1 Which variable has the effect shown in Figure 13.11 on rate of reaction? [1 mark]

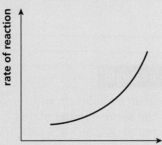

Figure 13.11

A concentration of solution B pressure of gases C surface area of solids D temperature

2 Hydrogen reacts with nitrogen to form ammonia. Iron acts as a catalyst for this reaction. Which of the graphs in Figure 13.12 shows how the mass of iron varies during the reaction? [1 mark]

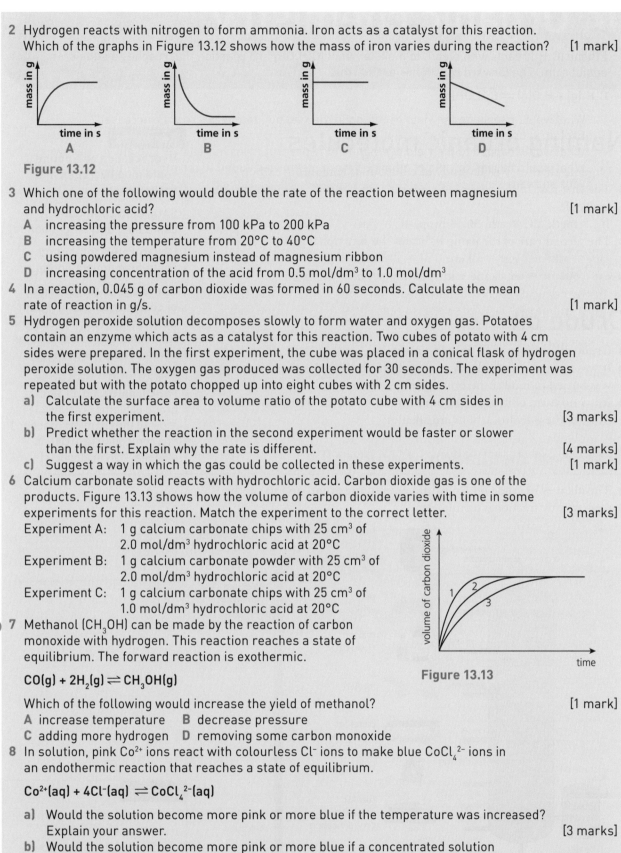

Figure 13.12

3 Which one of the following would double the rate of the reaction between magnesium and hydrochloric acid? [1 mark]
   A increasing the pressure from 100 kPa to 200 kPa
   B increasing the temperature from 20°C to 40°C
   C using powdered magnesium instead of magnesium ribbon
   D increasing concentration of the acid from 0.5 mol/dm³ to 1.0 mol/dm³

4 In a reaction, 0.045 g of carbon dioxide was formed in 60 seconds. Calculate the mean rate of reaction in g/s. [1 mark]

5 Hydrogen peroxide solution decomposes slowly to form water and oxygen gas. Potatoes contain an enzyme which acts as a catalyst for this reaction. Two cubes of potato with 4 cm sides were prepared. In the first experiment, the cube was placed in a conical flask of hydrogen peroxide solution. The oxygen gas produced was collected for 30 seconds. The experiment was repeated but with the potato chopped up into eight cubes with 2 cm sides.
   a) Calculate the surface area to volume ratio of the potato cube with 4 cm sides in the first experiment. [3 marks]
   b) Predict whether the reaction in the second experiment would be faster or slower than the first. Explain why the rate is different. [4 marks]
   c) Suggest a way in which the gas could be collected in these experiments. [1 mark]

6 Calcium carbonate solid reacts with hydrochloric acid. Carbon dioxide gas is one of the products. Figure 13.13 shows how the volume of carbon dioxide varies with time in some experiments for this reaction. Match the experiment to the correct letter. [3 marks]
   Experiment A: 1 g calcium carbonate chips with 25 cm³ of 2.0 mol/dm³ hydrochloric acid at 20°C
   Experiment B: 1 g calcium carbonate powder with 25 cm³ of 2.0 mol/dm³ hydrochloric acid at 20°C
   Experiment C: 1 g calcium carbonate chips with 25 cm³ of 1.0 mol/dm³ hydrochloric acid at 20°C

Figure 13.13

7 Methanol (CH_3OH) can be made by the reaction of carbon monoxide with hydrogen. This reaction reaches a state of equilibrium. The forward reaction is exothermic.

$$CO(g) + 2H_2(g) \rightleftharpoons CH_3OH(g)$$

Which of the following would increase the yield of methanol? [1 mark]
   A increase temperature   B decrease pressure
   C adding more hydrogen   D removing some carbon monoxide

8 In solution, pink Co^{2+} ions react with colourless Cl^- ions to make blue $CoCl_4^{2-}$ ions in an endothermic reaction that reaches a state of equilibrium.

$$Co^{2+}(aq) + 4Cl^-(aq) \rightleftharpoons CoCl_4^{2-}(aq)$$

   a) Would the solution become more pink or more blue if the temperature was increased? Explain your answer. [3 marks]
   b) Would the solution become more pink or more blue if a concentrated solution containing Cl^- ions was added? Explain your answer. [3 marks]

Answers and quick quiz 13 online

ONLINE

14 Organic chemistry

Naming organic molecules

- Organic chemistry is the study of compounds containing carbon.
- The first part of the name indicates the number of C atoms in the molecule.
  1C = meth, 2C = eth, 3C = prop, 4C = but
- The second part of the name indicates the functional group.
  –ane = alkane, –ene = alkene, –ol = alcohol
- e.g. butane = an alkane with 4 carbon atoms

Crude oil

- Crude oil is a fossil fuel found in rocks.
- It was formed over millions of years from the remains of plankton that was buried in mud at the bottom of the oceans.
- It is a mixture of many compounds, most of which are hydrocarbons. Most of these hydrocarbons are alkanes.

Fractional distillation of crude oil

- The alkanes in crude oil have different boiling points and so can be separated by fractional distillation.

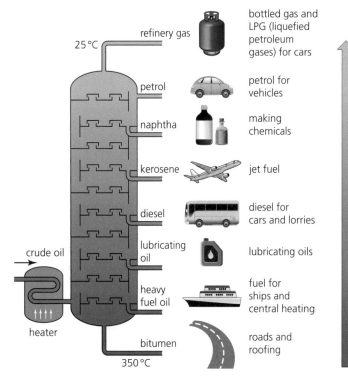

Figure 14.1 Fractional distillation of crude oil

- The crude oil is vaporised and passed into a column that is hot at the bottom and cool at the top.
- The vapour rises up the tower, and the different alkanes cool and condense at different heights.
- The bigger the molecule, the higher the boiling point and the lower down the tower it condenses.
- The process separates the alkanes into fractions, each containing a mixture of alkanes with similar boiling points.
- The fractions are processed to produce many fuels (e.g. petrol, diesel) and feedstock to make many key products (e.g. solvents, polymers, detergents).

Alkanes

- Alkanes are a homologous series of saturated hydrocarbons.
- Alkanes have the general formula C_nH_{2n+2}.

Table 14.1

	methane	ethane	propane	butane
molecular formula	CH_4	C_2H_6	C_3H_8	C_4H_{10}
displayed formula	H | H—C—H | H	H H | | H—C—C—H | | H H	H H H | | | H—C—C—C—H | | | H H H	H H H H | | | | H—C—C—C—C—H | | | | H H H H

Properties of alkanes

REVISED ☐

- The bigger the alkane molecule:
  ○ the higher the boiling point
  ○ the less flammable (i.e. catches fire less easily)
  ○ the more viscous (i.e. flows less easily).

The use of alkanes as fuels

REVISED ☐

- The main use of alkanes is as fuels. They burn well and release a lot of heat energy.
- Complete combustion takes place in a good supply of oxygen; this forms carbon dioxide and water.

e.g. propane + oxygen → carbon dioxide + water

$$C_3H_8 + 5O_2 \rightarrow 3CO_2 + 4H_2O$$

Exam tip

- The fastest way to balance equations for the complete combustion of alkanes is:
  - the number of CO_2 molecules equals the number of C atoms in the alkane
  - the number of H_2O molecules equals half the number of H atoms in the alkane
  - count the total number of O atoms in the CO_2 and the H_2O; the number of O_2 molecules is half this number.

Cracking alkanes

- Shorter alkanes are in very high demand as fuels and it is hard to meet this demand.
- Larger alkanes are in less demand as fuels as they are harder to ignite and are more viscous.
- Larger alkanes can be broken down by **cracking** to form smaller alkanes (used as fuels) and alkenes (used to make polymers and other chemicals).
- Cracking is a thermal decomposition reaction.

long alkanes → shorter alkanes + alkenes

Table 14.2

type of cracking	catalytic cracking	steam cracking
method	vaporise the alkanes and pass them over a hot catalyst	vaporise the alkanes, mix them with steam, and heat to high temperature

Now test yourself

1 Hexane is an alkane with 6 carbon atoms.
  (a) Give the molecular formula of hexane.
  (b) Draw the displayed formula of hexane.
  (c) Hexane is a saturated hydrocarbon. Explain what the terms 'saturated' and 'hydrocarbon' mean.
2 Decane has the formula $C_{10}H_{22}$; pentane has the formula C_5H_{12}. Which of these two alkanes:
  (a) has the higher boiling point?
  (b) is more viscous?
  (c) is more flammable?
  (d) would collect higher up a fractional distillation column?
3 Cracking icosane ($C_{20}H_{42}$) forms decane ($C_{10}H_{22}$), propene (C_3H_6) and butene (C_4H_8).
  (a) Describe one way in which cracking can be done.
  (b) Write a balanced equation for this reaction.
4 Write a balanced equation for the complete combustion of pentane (C_5H_{12}).

Answers online

Alkenes

Exam tip

You do not need to learn the names, formulae or structures of individual alkenes, but you should be able to recognise them from their names, formulae or structure.

- Alkenes are a homologous series of unsaturated hydrocarbons.
- The functional group is a C=C double bond.
- Alkenes have the general formula C_nH_{2n}.
- Alkenes react with bromine water, turning it from orange–yellow to colourless. This is a way to test for alkenes.

Table 14.3

	ethene	propene	butene	pentene
molecular formula	C_2H_4	C_3H_6	C_4H_8	C_5H_{10}
displayed formula	H–C=C–H (with H on each)	H–C–C=C–H	H–C–C–C=C–H	H–C–C–C–C=C–H

Summary

- Crude oil is a mixture of hydrocarbons, which are mainly alkanes.
- Alkanes are saturated hydrocarbons with the general formula C_nH_{2n+2}.
- The alkanes in crude oil have different boiling points and so can be separated into fractions by fractional distillation.
- The bigger the alkane, the higher the boiling point, the less flammable and the more viscous it is.

- Alkanes are very good fuels, but shorter alkanes are in more demand as fuels.
- Longer alkanes are broken down into shorter alkanes and alkenes by cracking.
- Alkenes are unsaturated hydrocarbons.
- Alkenes are reactive and turn bromine water colourless. This is used as a test for alkenes.

Exam practice

1 Which of the following is the molecular formula of propane? [1 mark]
  A C_3H_6 B C_3H_8 C C_5H_{10} D C_5H_{12}

2 Fractional distillation is used to separate the alkanes in crude oil into fractions. Which property does this separation technique depend on? [1 mark]
  A boiling points   B flammability   C melting points   D viscosity

3 Undecane is an alkane with the formula $C_{11}H_{24}$. Some is burned as a fuel and some is cracked.
  (a) Balance the equation for the complete combustion of undecane. [1 mark]

$$C_{11}H_{24} + O_2 \rightarrow CO_2 + H_2O$$

  (b) Balance the equation for the cracking of undecane. [1 mark]

$$C_{11}H_{24} \rightarrow C_4H_{10} + C_2H_4 + C_3H_6$$

  (c) The cracking reaction produces butane (C_4H_{10}). Which is more flammable, butane or undecane? [1 mark]
  (d) Describe one way in which cracking can be done and explain clearly why it is done. [6 marks]

4 What would you see if a few drops of bromine water were added to a boiling tube of ethene gas? [1 mark]

Answers and quick quiz 14 online

ONLINE

15 Chemical analysis

Purity, formulations and chromatography

Pure substances and mixtures

REVISED

Table 15.1

	pure substance	mixture
what it is	• a single substance (which could be an element or compound)	• more than one substance
melting and boiling points	• melt at a specific temperature • boil at a specific temperature	• melt and boil over a range of temperatures

Formulations

REVISED

- A formulation is a mixture of substances that has been designed as a useful product, e.g. alloys, cleaning agents, fertilisers, foods, fuels, medicines and paints.
- In a formulation, each chemical is present in a carefully controlled amount for a specific purpose. For example, in a headache tablet, ingredients include specific amounts of: the medicine(s) to relieve the pain; binders to hold the ingredients together; and a coating to make the tablet easier to swallow.

Now test yourself

TESTED

1 Which of the following are mixtures? Tap water, air, mineral water, gold alloy, oxygen, diamond.
2 Kerosene is a fraction of crude oil used as jet fuel. It boils over the temperature range 205–260°C. Is kerosene a pure substance or a mixture? Explain your answer.
3 Toothpaste is a formulation. Explain what a formulation is.

Answers online

Chromatography

REVISED

- Chromatography is a very useful technique that can be used to separate and analyse mixtures.
- There are many types of chromatography including paper, thin layer, column and gas chromatography.
- All forms of chromatography involve two phases – a stationary and a mobile phase.
- In paper chromatography, the stationary phase is paper and the mobile phase is a solvent.
- The separation works because the substances in the mixture have a different relative attraction to the stationary phase (paper) and mobile phase (solvent) compared to each other. This means that they travel at different speeds along the paper.
- The R_f value for each substance is the ratio of the distance moved by the substance to the distance moved by the solvent:

$$R_f = \frac{\text{distance moved by substance}}{\text{distance moved by solvent}}$$

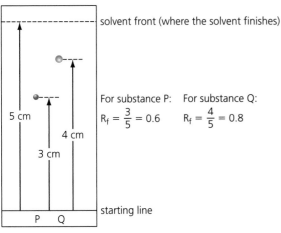

Figure 15.1 Finding R_f values

solvent front (where the solvent finishes)

For substance P: For substance Q:

$R_f = \dfrac{3}{5} = 0.6$ $R_f = \dfrac{4}{5} = 0.8$

5 cm

4 cm

3 cm

starting line

P   Q

Now test yourself

TESTED

4 Food colouring K was analysed by paper chromatography and compared to dyes W, X, Y and Z.

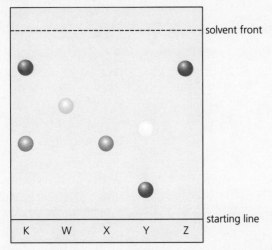

solvent front

starting line

K   W   X   Y   Z

Figure 15.2

(a) Calculate the R_f values for the substances in K.
(b) Which substances are in K?

Answers online

Required practical 12

Paper chromatography

Paper chromatography can be used to separate and tell the difference between coloured substances.

AIM: To analyse some food dyes by paper chromatography.

Table 15.2

Instructions	Comments
1 A pencil line is drawn on a piece of chromatography paper near the bottom.	● Pencil is used as it will not dissolve in the solvent.
2 Small amounts of the substances being analysed are placed as spots on the pencil line using capillary tubes.	● Once the drop has dried, more of the dye being analysed can be added onto the drop. This means that there is lots of the substance being analysed without the spot being too big.

Table 15.2 *continued*

3	The chromatography paper is hung in a beaker of the solvent. The pencil line and spots must be above the level of the solvent.	• The line and spots must be above the level of the solvent so that the spots do not dissolve in the solvent. • Water is often used as the solvent, but other solvents may be used if they give better separation for some samples.
4	Over the next few minutes, the solvent soaks up the paper.	
5	When the solvent is near the top, the paper is removed from the solvent. The level that the solvent reaches should be marked on the paper.	• The final level the solvent reached (sometimes called the solvent front) is marked so that the R_f values for each spot can be calculated.
6	The paper is left to dry.	

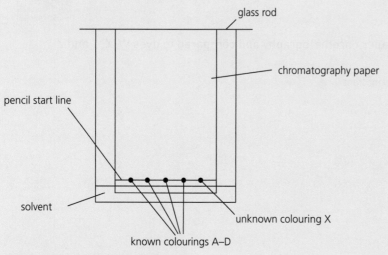

Figure 15.3 Experiment set-up

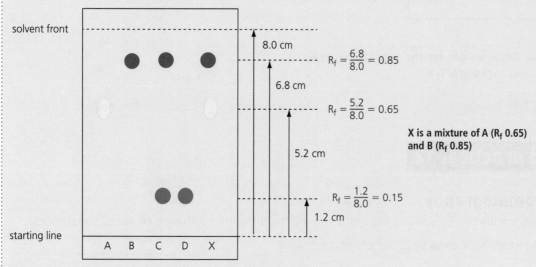

$$R_f = \frac{6.8}{8.0} = 0.85$$

$$R_f = \frac{5.2}{8.0} = 0.65$$

X is a mixture of A (R_f 0.65) and B (R_f 0.85)

$$R_f = \frac{1.2}{8.0} = 0.15$$

Figure 15.4 Analysis of dye X compared to dyes A–D

Answers and quick quizzes at **www.hoddereducation.co.uk/myrevisionnotesdownloads**

Summary

- A pure substance is a single element or compound. Pure substances each melt and boil at specific temperatures.
- A mixture contains two or more substances. Mixtures melt and boil over a range of temperatures.
- A formulation is a mixture of substances in specific amounts that has been designed as a useful product (e.g. medicines, fertilisers).
- Chromatography can be used to separate and analyse mixtures. Each type has a stationary phase and a mobile phase. Separation works because the substances have a different relative affinity for these two phases.
- Common gases can be tested:
  - carbon dioxide gas (CO_2) turns limewater cloudy
  - chlorine gas (Cl_2) bleaches damp litmus paper
  - hydrogen gas (H_2) makes a squeaky pop with a burning splint
  - oxygen gas (O_2) relights a glowing splint.

Exam practice

1 Food colouring S is a formulation. It was analysed by paper chromatography and compared to substances 1–6.
  (a) What is a formulation? [1 mark]
  (b) Which of the substances 1–6 are could be pure substances? How can you tell? [2 marks]
  (c) Which substances are in S? [1 mark]
  (d) What is the R_f values of the substances in S? Give your answers to 2 significant figures. [2 marks]
  (e) Why was the starting line drawn in pencil? [1 mark]
  (f) Explain why the substances separate in paper chromatography. [3 marks]

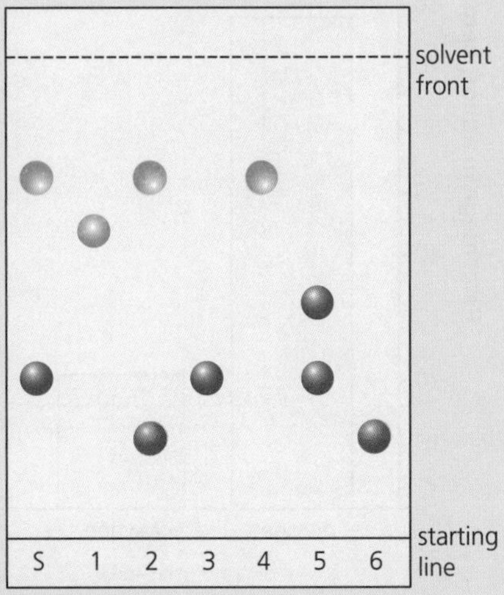

Figure 15.5

2 When sodium hydroxide solution is added to a solution of W, a white precipitate is formed that re-dissolves when more sodium hydroxide is added. What is the positive ion in W? [1 mark]

   A Al^{3+} B Ca^{2+} C Cu^{2+} D Mg^{2+}

3 In a test to identify the gas formed in an electrolysis experiment, the gas bleached a piece of red litmus paper. Identify the gas. [1 mark]

   A carbon dioxide   B chlorine   C hydrogen   D oxygen

Answers and quick quiz 15 online

ONLINE

16 Chemistry of the atmosphere

The Earth's atmosphere

The atmosphere today

- For about the last 200 million years, the atmosphere has contained:
  - about four-fifths (80%) nitrogen (N_2)
  - about one-fifth (20%) oxygen (O_2)
  - small proportions of other gases (e.g. argon, carbon dioxide, water vapour).

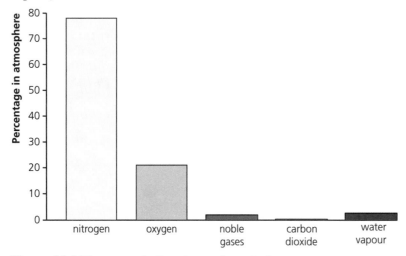

Figure 16.1 The gases in the atmosphere today

The early atmosphere of the Earth

- The Earth is about 4.6 billion years old.
- There are many theories about how the Earth's atmosphere has changed in that time.
- One theory is that during the first billion years there was lots of volcanic activity which released the gases that formed the early atmosphere. These volcanoes may have released:
  - carbon dioxide (CO_2) – this was probably the main gas in the early atmosphere
  - nitrogen (N_2) – this may have built up over time
  - methane (CH_4) – small amounts
  - ammonia (NH_3) – small amounts
  - water vapour (H_2O) – as the hot earth cooled down this may have condensed to form the oceans.
- The young Earth may have had an atmosphere like that of Mars and Venus today whose atmospheres are mainly carbon dioxide.

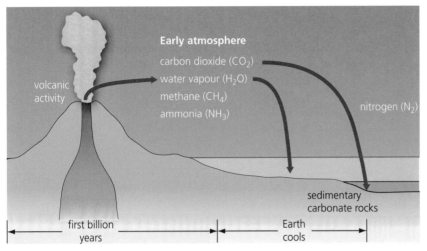

Figure 16.2 Volcanoes released the gases that formed the early atmosphere

How the atmosphere changed

REVISED

Where did the oxygen come from?

- There was probably little or no oxygen in the atmosphere when the Earth was young.
- Life evolved on Earth and it may be that the oxygen was formed by photosynthesis in living creatures, such as algae at first and later by plants.

$$6CO_2 + 6H_2O \rightarrow 6O_2 + C_6H_{12}O_6$$

carbon dioxide + water → oxygen + glucose

> **Typical mistake**
>
> Students often mix up photosynthesis and respiration. Photosynthesis uses up carbon dioxide and makes oxygen. The opposite happens in respiration.

Where did the carbon dioxide go?

- Most of the carbon dioxide has been removed from the atmosphere (there is now only about 0.04%)
- Most of the carbon atoms from the carbon dioxide have become locked up over millions of years in fossil fuels (e.g. coal, oil, natural gas) or in compounds in sedimentary rocks as carbonates (e.g. as calcium carbonate in limestone).

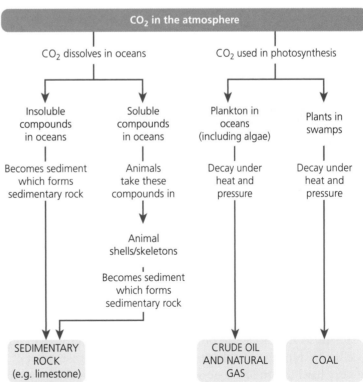

Figure 16.3 Ways in which carbon dioxide has been removed from the atmosphere

Now test yourself

1 Which of the following is thought to have been the main gas in the Earth's early atmosphere?
  A ammonia   B carbon dioxide   C methane   D oxygen
2 Describe how intense volcanic activity on the young Earth may have led to the formation of the oceans.
3 Describe how oxygen in the atmosphere was formed.
4 Where is most of the carbon dioxide that was in the early atmosphere now?

Answers online

Greenhouse gases

What are greenhouse gases?

- Greenhouse gases include carbon dioxide (CO_2), methane (CH_4) and water vapour (H_2O).
- Greenhouse gases do not absorb the sun's radiation but do absorb the Earth's radiation. This is because the Earth's infrared radiation has a longer wavelength than the sun's.
- These greenhouse gases help to keep the Earth warm allowing life on our planet.

radiation from Sun (sunlight)
radiation from Earth
some of Earth's radiation absorbed by greenhouse gases

Figure 16.4

Increasing levels of greenhouse gases in the atmosphere

- The total amount of greenhouse gases in the atmosphere is increasing due to human activities.

Table 16.1

greenhouse gas	is the amount increasing?	reason
carbon dioxide (CO_2)	increasing	• large scale burning of fossil fuels • deforestation (massive areas of forest are being cut down)
methane (CH_4)	increasing	• increased animal farming (produced by digestion in cattle) • increased amount of rubbish in landfill (as rubbish rots it gives off methane)
water vapour (H_2O)	no overall increase	• this varies due to weather conditions • human activities have no impact on amount of H_2O in the air

Scientific views on global warming and climate change

- The amount of greenhouse gases in the atmosphere is increasing and the Earth is getting warmer.
- Most scientists think that the Earth is getting warmer due to the increase in the amount of greenhouse gases in the atmosphere.
- This is based on peer-reviewed evidence where research is checked by other scientists before it is published.
- Scientists use models to predict how different factors will affect the climate. However, as there are so many factors that affect climate, it is very hard to produce a good model. This means that over-simplified models are used and these can lead to misleading reports in the media based on only part of the evidence.

Answers and quick quizzes at **www.hoddereducation.co.uk/myrevisionnotesdownloads**

The impact of global warming and climate change

- It is hard to predict the effects of global warming. Some possible effects are shown in Table 16.2.

Table 16.2

effect	details
sea level rise	<ul><li>due to ice caps melting and the thermal expansion of water in the oceans</li><li>likely to lead to some regions being flooded or submerged</li></ul>
storms and rainfall	<ul><li>there may be more frequent and severe storms</li><li>the amount, timing and distribution of rainfall may change (some areas may get more rain but others less)</li></ul>
wildlife	<ul><li>species may migrate north as the temperature rises</li><li>plants may flower earlier</li><li>animals may come out of hibernation earlier</li></ul>
food production	<ul><li>the ability of some regions to produce enough food may be affected by changes in temperature and rainfall</li></ul>

Carbon footprint

- Carbon footprint is the total amount of carbon dioxide and other greenhouse gases emitted over the full life cycle of a product, service or event.
- For example, the carbon footprint for the use of a plastic bag would include carbon dioxide being released from energy generation when crude oil is drilled, separated by fractional distillation, cracked to make alkenes, made into polymers, made into plastic bags, transported to shops, transported to waste disposal sites, etc.
- It is hard to produce an exact figure for carbon footprint, but it is still a useful guide and can identify products, services or events with high carbon footprints.
- Some examples of the many ways to reduce carbon footprints are shown in Table 16.3.

Table 16.3

way to reduce carbon footprint	more detail
more use of renewable energy sources	<ul><li>e.g. wind power, tidal power, solar power</li></ul>
more use of nuclear power	<ul><li>nuclear fuels do not release greenhouse gases</li></ul>
more use of biofuels	<ul><li>biofuels are fuels made from crops (e.g. biodiesel, ethanol)</li><li>they are carbon neutral, i.e. the crops absorb the same amount of CO_2 when they grow as the fuels release when they burn</li></ul>
increase energy efficiency	<ul><li>there are many examples of this, e.g.<ul><li>increasing insulation in homes</li><li>use LED light bulbs instead of halogen bulbs</li><li>switch off electrical devices instead of leaving them on standby</li></ul></li></ul>
carbon capture and storage	<ul><li>collect the CO_2 given off when fuels burn and store it underground</li></ul>

- There are some problems that make the reduction in greenhouse gases difficult, which include:
  - **Scientific disagreement** – not all scientists agree that emissions of greenhouse gases cause climate change.
  - **Cost** – many methods of reducing carbon footprints are more expensive than doing nothing.

○ **Lack of international agreement** – some countries have not agreed to reduce their carbon footprint.
○ **Population and lifestyle changes** – as the world's population increases and lifestyle develops, there is an increasing demand for energy.
○ **Education** – many people are confused and do not understand what needs to be done and why.

5 Carbon dioxide and methane are greenhouse gases. Explain why they absorb the Earth's infrared radiation but not the sun's.
6 List three likely consequences of global warming.
7 Suggest three ways in which a school could reduce its carbon footprint.

Answers online

Common pollutants in the atmosphere   REVISED

● The combustion of fuels is a major source of atmospheric pollutants.
● Fossil fuels contain carbon and/or hydrogen and may also contain some sulfur.
● The diagram shows substances formed when fossil fuels are burned, the problems they may cause and ways to reduce these problems.

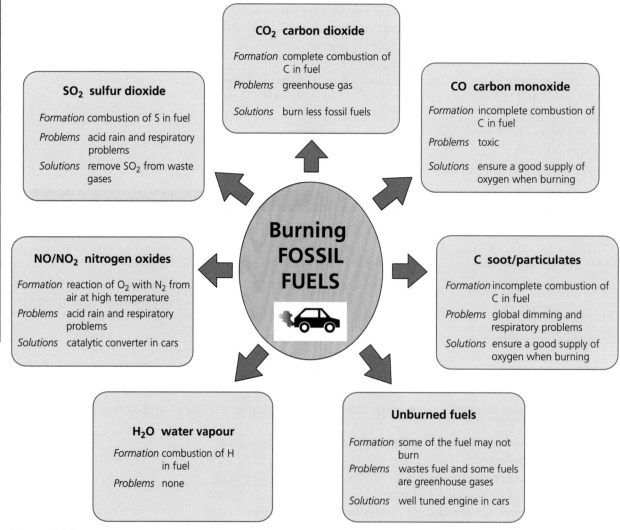

Figure 16.5

Notes about the diagram

- Complete combustion takes place when a substance burns in a good supply of oxygen; incomplete combustion takes place when a substance burns in a poor supply of oxygen.
- Carbon monoxide is particularly dangerous to humans as it is colourless and has no smell, but is toxic.
- Carbon particulates cause global dimming; this is when the small soot particles block out some sunlight.

Now test yourself

TESTED

8 Name two gases formed when fossil fuels burn that cause acid rain.
9 Explain why carbon monoxide may be formed when fossil fuels are burned and what problem it can cause.

Answers online

Summary

- The Earth's atmosphere today is about four-fifths nitrogen and one-fifth oxygen, with small amounts of argon, water vapour and carbon dioxide.
- The Earth's early atmosphere may have been like that of Mars and Venus today, containing mainly carbon dioxide.
- The gases in the early atmosphere may have been released from volcanoes.
- There was probably little or no oxygen in the early atmosphere. Oxygen was formed by photosynthesis in the first algae and then plants as well.
- Most carbon dioxide has been removed from the atmosphere and is now locked up in sedimentary rocks or fossil fuels.
- Carbon dioxide, methane and water vapour are greenhouse gases that absorb the Earth's infrared radiation trapping heat in the atmosphere.
- The amount of greenhouse gases is increasing due to large scale use of fossil fuels, deforestation, increasing farming and use of landfill.
- The earth is warming up (global warming) and most scientists believe it is due to the increasing amount of greenhouse gases in the atmosphere.

- The effects of global warming are hard to predict, but they are likely to lead to sea levels rising, climate change and impacts on wildlife and food production.
- Carbon footprint is the total amount of carbon dioxide and other greenhouse gases emitted over the full life cycle of a product, service or event.
- There are many ways to reduce carbon footprints, including more use of renewable energy sources and improved energy efficiency.
- There are barriers to reducing carbon footprint, which include scientific disagreement, cost, rising world population and lack of international agreement.
- When fossil fuels are burned, the following pollutants are formed:
  - carbon dioxide (CO_2), which is a greenhouse gas
  - carbon monoxide (CO), which is toxic, from incomplete combustion
  - carbon particulates (soot, C), which causes respiratory problems and global dimming
  - sulfur dioxide (SO_2), which causes respiratory problems and acid rain
  - nitrogen oxides (NO and NO_2), which cause respiratory problems and acid rain.

Exam practice

1 Identify the planet whose atmosphere at present is thought to be most like that of the Earth when it was young.
   A Mercury   B Mars   C Jupiter   D Neptune   [1 mark]
2 What figure best represents the percentage of oxygen in the atmosphere today?
   A 20%   B 25%   C 75%   D 80%   [1 mark]

→

3 Balance the equation for photosynthesis. [1 mark]

$CO_2 + H_2O \rightarrow C_6H_{12}O_6 + O_2$

4 Describe how carbon dioxide from the early atmosphere became locked up in natural gas and crude oil. [5 marks]

5 Carbon dioxide and methane are greenhouse gases. Explain what a greenhouse gas is and why the amount of each gas is increasing in the atmosphere. [6 marks]

6 The graph in Figure 9.6 shows the emissions of total greenhouse gases and carbon dioxide in the UK from 1990–2015. Describe the trends shown by this graph and suggest reasons for these trends. [4 marks]

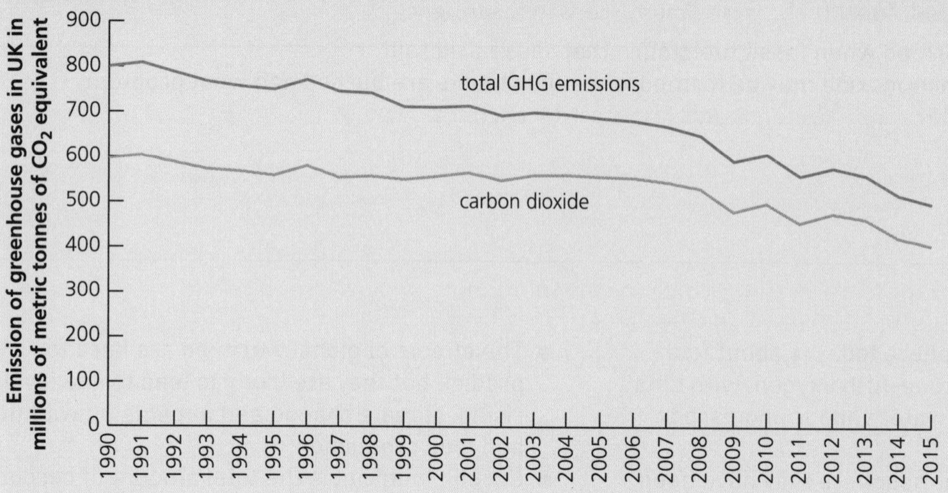

Figure 16.6

7 Match each of the following pollutants to the problem it causes. [2 marks]

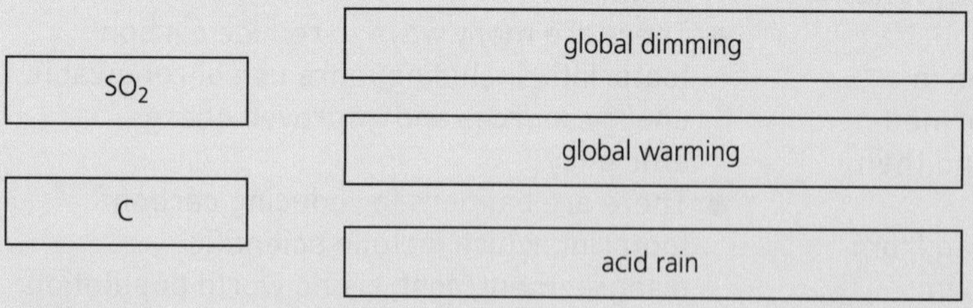

Figure 16.7

8 Explain how each of the following gases are formed when some fossil fuels are burned.
(a) sulfur dioxide (SO_2)
(b) nitrogen oxide (NO) [3 marks]

Answers and quick quiz 16 online

ONLINE

17 Using the Earth's resources

The Earth's resources

- Everything we need for life is provided by the Earth's resources, e.g. rocks in the ground, fossil fuels, fresh water, sea water, sunlight, wind, etc.
- We make many substances, such as medicines, metals, polymers and some clothing fibres from these natural resources by chemical reactions.

Sustainable development

REVISED

- Many of the Earth's resources are finite which means we cannot replace them once we have used them (e.g. crude oil).
- Other resources are renewable, which means we can replace them (e.g. biofuels).
- Sustainable development is where we use resources to meet the needs of people today without preventing people in the future from meeting their needs.
- It is important that we look to meet our needs today in a sustainable way.
- For example, we could use renewable energy sources (e.g. solar, wind) instead of using fossil fuels.

Re-use and recycling

REVISED

- Some products can be re-used. For example, glass milk bottles can be washed and used over and over again.
- Some products can be recycled. Most metals, glass and plastics can be recycled by melting down and being made into new products. The different types of metal, glass and plastic have to be separated first.
- If we re-use or recycle products, it will:
  ○ reduce the use of the Earth's limited resources
  ○ reduce energy consumption and the use of fuels
  ○ reduce the impact of manufacturing new materials on the environment (e.g. less mining and quarrying)
  ○ reduce the impact of waste disposal on the environment.

Life cycle assessment (LCA)

REVISED

- A life cycle assessment (LCA) reviews the impact of a product on the environment throughout its life. This includes:
  ○ use and sustainability of raw materials (including those for packaging)
  ○ use of energy at all stages
  ○ use of water at all stages
  ○ production and disposal of waste products at all stages
  ○ transport and distribution at all stages.
- Simple LCAs for a plastic and a paper shopping bag are shown in Table 17.1.

Table 17.1

		plastic shopping bags	paper shopping bags
raw materials	what they are	crude oil	trees
	sustainability	not sustainable – crude oil cannot be replaced	sustainable – more trees can be planted (but take a long time to grow)
	obtaining raw materials	extracting crude oil uses lots of energy	habitats are destroyed as trees are cut down
	transporting raw materials	transport of crude oil uses up fuels and causes some pollution; potential damage to environment from spillages	transport of logs uses up fuels and causes some pollution
manufacture of bags		much energy used to separate crude oil into fractions, for cracking and for polymerisation	uses a lot of water; uses some harmful chemicals (leaks would damage the environment)
use of bags	transport to where used	transport of bags uses up fuels and causes some pollution	
disposal options	landfill	does not rot and so remains in ground for many years	rots releasing greenhouse gases including methane
	incinerator	gives off the greenhouse gas carbon dioxide when burned	
	recycled	transport of bags uses up fuels and causes some pollution; melting of plastic uses energy	transport of bags uses up fuels and causes some pollution; relatively easy to recycle

● Many factors in a LCA can be easy to quantify, but others (e.g. effect of pollutants) are harder. This means that an LCA may not be completely objective.

Now test yourself

TESTED

1 What is the difference between a finite and a renewable resource. Give one example of each.
2 Give two key reasons why the disposal of plastic bags in landfill sites is not sustainable.
3 Some glass bottles are recycled and some are re-used. Explain the difference.

Answers online

The use of water

REVISED

Table 17.2 Types of water

type	description	contents	
		dissolved substances	microbes
pure water	water that contains only water molecules and nothing else	✗	✗
potable water	water that is safe to drink	✓ low levels	✗ (or very low levels)
fresh water	water found in places, such as lakes, rivers, the ice caps, glaciers and underground rocks and streams	✓ low levels	✓ (very low from some sources)
ground water	fresh water found in underground streams and porous rocks (aquifers)	✓ low levels	✓
sea water	water in the seas and oceans	✓ high levels	✓
waste water	used water from homes, industry and agriculture	✓ high levels	✓

- Water is essential for life.
- **Potable** water is water that is safe to drink. It is not pure as it contains some dissolved substances but at low levels which are safe.

> **Potable** water that is fit to drink

Water treatment

- In the UK, we produce potable water from fresh water. The two main stages in this process are filtration and sterilisation.

Table 17.3

stage	what it does	details
filtration	removes solids	<ul><li>the water is passed through filter beds</li><li>these filter beds are usually made of sand</li><li>as the water flows through the sand, any solids in the water are removed</li></ul>
sterilisation	kills microbes	<ul><li>the most common way is to use small amounts of chlorine</li><li>however, the microbes can also be killed by treating the water with ozone or by passing ultraviolet light through the water</li></ul>

- In some parts of the world, there is little fresh water but lots of sea water.
- Potable water can be made from sea water by desalination, which is done by distillation or reverse osmosis.
- Both methods are expensive as they require a lot of energy.

Table 17.4

method	distillation	reverse osmosis
diagram		
notes	<ul><li>sea water is heated so it boils</li><li>as the steam leaves the dissolved substances are left behind</li><li>the steam is then cooled and condensed</li><li>the process is expensive due to the high cost of heat energy to boil the water</li></ul>	<ul><li>sea water is passed through a semi-permeable membrane under pressure</li><li>the water molecules pass through the membrane but many dissolved substances cannot</li><li>the process is expensive due to the high cost of energy needed to provide the high pressure</li></ul>

Waste water treatment

- Large amounts of waste water are produced.
- Waste water has to be treated before it can be returned to the environment.

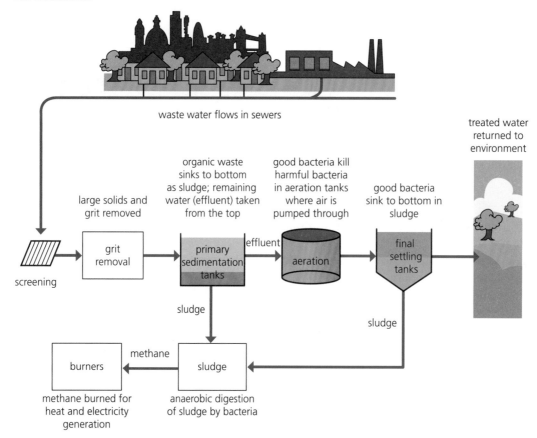

Figure 17.1 **Waste water treatment**

Now test yourself

4 (a) What is potable water?
   (b) Give one way in which potable water is different to pure water.
5 Identify three ways in which water can sterilised to make it fit for drinking.
6 Identify two methods used to desalinate sea water in countries that have little fresh water.
7 Why must waste water be treated before it can be returned to the environment?

Answers online

Required practical 13

Analysis and purification of water

AIM: To distil sea water and to analyse water from various sources.

Table 17.5

PART 1 – Distillation of sea water		● boil some sea water in a conical flask ● cool the water vapour given off with an ice bath and collect the distilled water in a test tube. 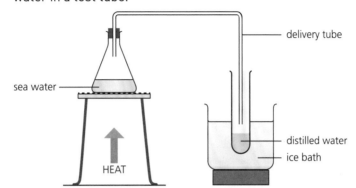
PART 2 – Analysis of water samples	finding mass of dissolved solids	● find the mass of a dry watch glass ● place 4 cm³ of water on the watch glass and find the mass again ● heat the watch glass on a water bath to evaporate all the water ● leave the watch glass to cool ● dry the underside of the watch glass ● find the mass of the watch glass which will may have some residue from the evaporated water on it. 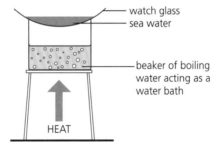
	measuring pH	● use universal indicator paper and a colour chart to test the pH

Table 17.6

type of water	sea water	mineral water	rain water	distilled water
mass of residue on watch glass (g)	0.14	0.08	0.00	0.00
pH of water sample	8	6	5	7

Notes
● In terms of dissolved solids, sea water contains the most, followed by mineral water. Rain water and distilled water do not contain any dissolved solids.
● In terms of pH, sea water is alkaline and distilled water is neutral. Mineral water and rain water are acidic, with rain water being more acidic.

Metals

ⒽExtraction of copper

- Metals are extracted from compounds in rocks called ores.
- The Earth's resources of metal ores are limited.
- Copper ores are very limited and new methods of extracting copper from low-grade ores have been developed.
- For example, phytomining uses plants to absorb metal compounds from the ground. This avoids, for example, traditional mining methods of digging, moving and disposing of large amounts of rock.

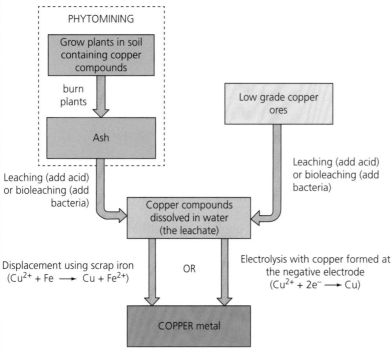

Figure 17.2 **Phytomining**

Ⓗ 8 Copper can be extracted from the ash of plants grown in soil rich in copper compounds.
   (a) What is the name of this process?
   (b) A leachate solution containing water soluble copper compounds is formed from the ash. State two ways in which copper metal can be extracted from this leachate?

Summary

- Some of the Earth's resources are renewable, but many are finite and cannot be replaced once used.
- Sustainable development is where we use resources to meet the needs of people today without preventing people in the future from meeting their needs.
- The re-use or recycling of products improves sustainability.

- A life cycle assessment (LCA) is one way to evaluate the impact of a product on the environment throughout its life, including use of resources.
- Potable water is water that is fit to drink.
- The production of potable water from fresh water includes filtration and sterilisation.
- Potable water can be produced by desalination of sea water. This is done by distillation or reverse osmosis.

● Waste water is treated before it can be returned to the environment.
● Metals are produced from compounds in rocks called ores.

H ● Methods are being developed to produce metals from low-grade ores. For example, phytomining uses plants to absorb metal compounds from the ground. The plants are then burned and the metal extracted from compounds in the ash.

Exam practice

H **1** Which one the following increases the equilibrium yield of ammonia in the Haber process? [1 mark]
   A increase pressure   **B** increase temperature
   C use an iron catalyst   **D** increase the surface area of the iron catalyst

2 Match the following types of water to the correct descriptions. [2 marks]

	water that is fit to drink
fresh water	
	water that contains water molecules and nothing else
potable water	
	water found in lakes, rivers, ice caps, glaciers

3 Which of the following is the most sustainable way to dispose of used plastic bags? [1 mark]
   A bury in landfill   **B** re-use   **C** recycle   **D** incinerate

H **4** Copper can be extracted from the copper(II) compounds, such as copper(II) sulfate, in the leachate solutions formed from treatment of plant ash from phytomining.
   (a) One way to extract copper from copper(II) sulfate is by displacement using iron. Explain why iron can be used in this way and write a word equation for this reaction. [2 marks]
   (b) Copper can also be extracted from copper(II) sulfate by electrolysis using graphite electrodes. At which electrode is copper formed and write a half equation for the reaction at this electrode. [2 marks]
   (c) Explain why both these extraction processes involve reduction. [1 mark]

5 Waste water must be treated before it can be returned to the environment.
   Describe process X which produces methane gas from sludge formed in the process. [3 marks]

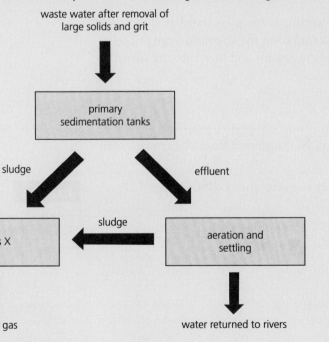

waste water after removal of
large solids and grit

↓

primary
sedimentation tanks

sludge ↙   ↘ effluent

process X	← sludge — aeration and settling

↓   ↓

methane gas   water returned to rivers

Answers and quick quiz 17 online

ONLINE

18 Energy

Energy changes in a system, and the ways energy is stored before and after such changes

A system is an object or group of objects. There are changes in the way energy is stored when a system changes.

- **Throwing an object upwards**

  When you throw a ball upwards, just after the ball leaves your hand, it has a store of kinetic energy. By the time the ball reaches its highest point, the kinetic energy has been transferred to a gravitational potential energy store. Just before you catch the ball, it has a store of kinetic energy again.

- **A moving object hitting an obstacle**

  If you drop a lump of clay, it sticks to the ground. Just before the clay hits the ground, it has a store of kinetic energy. After it has hit the ground, the kinetic energy has all been transferred to thermal energy, which is stored in the clay or in the surroundings. We hear a sound when the clay hits the ground, but the sound is quickly dissipated as thermal energy in the surroundings.

- **Accelerating a car with a constant force**

  A force on the car does work to accelerate the car. There is a store of chemical energy in the car's petrol. As the petrol burns, it transfers energy into the kinetic energy store of the moving car and also into the thermal store of the surroundings.

- **Heating a resistor**

  When a current flows through a resistor, energy is transferred by electrical work. Energy is transferred from the chemical store in the battery into the thermal store of the resistor and then into the thermal store of the surroundings.

> **Exam tip**
>
> Energy can be transferred from one store to another.

Energy stores

REVISED

When a system changes, energy can be transferred from one energy store to another.

You will meet the following energy stores in your GCSE course:

- kinetic
- chemical
- gravitational potential
- magnetic
- elastic potential
- electrostatic
- thermal (or internal)
- nuclear.

> **Exam tip**
>
> Remember there are eight stores of energy.

Pathwaysto transfer energy

Energy can be transferred from one store to another by one of the following paths:
- heating
- work done by forces
- work done when a current flows
- electromagnetic radiation or mechanical radiation (shock waves and sound).

Example

A battery drives an electric motor that lifts a mass. Figure 18.1 shows the energy transfers that occur.

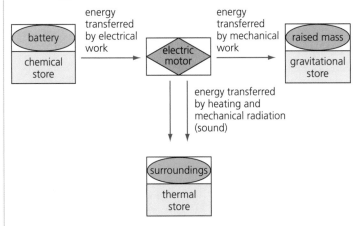

Figure 18.1

- Chemical energy is transferred from the battery by electrical work.
- The motor transfers gravitational potential energy to the mass by mechanical work.
- The motor also transfers energy to the thermal store of the surroundings by heating and mechanical radiation (sound).

At the start of the process, energy is stored in the battery; at the end, that energy is stored in the gravitational store of the mass and the thermal store of the surroundings.

The motor has a temporary store of kinetic energy when it turns.

Typical mistake

Do not describe light, sound and 'electrical energy' as types of energy or energy stores. Light, sound and electrical work are ways in which energy can be transferred from one store to another.

Counting the energy

Energy is a quantity that is measured in joules, J. Large quantities of energy are measured in kilojoules, kJ and megajoules, MJ.

$1\,kJ = 10^3\,J$

$1\,MJ = 10^6\,J$

The reason that energy is so important to us is that there is always the same energy at the end of a process as there was at the beginning.

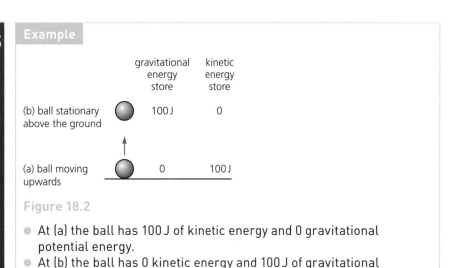

Figure 18.2

- At (a) the ball has 100 J of kinetic energy and 0 gravitational potential energy.
- At (b) the ball has 0 kinetic energy and 100 J of gravitational potential energy.

Energy is always conserved. The total amount of energy is the same at the beginning of a process as at the end.

The principle of conservation of energy

REVISED

The principle of conservation of energy states that the amount of energy always remains the same. There are various stores of energy. In a process, energy can be transferred from one store to another, but it cannot be created or destroyed.

Now test yourself

TESTED

1 Describe the energy stored in each of the following:
   (a) hot water in a bath
   (b) a car battery
   (c) a litre of petrol
   (d) a moving train
   (e) a golf ball in flight
   (f) a stretched spring.
2 Describe the changes involved in the way energy is stored when the following changes to a system occur. Explain where energy is stored at the beginning and end of each process.
   (a) A car is brought to a halt by applying its brakes.
   (b) An aeroplane accelerates along a runway and takes off.
   (c) A cell lights a torch lamp.
   (d) A hot cup of coffee cools down.
3 Figure 18.3 shows a ball falling from A to D. Write down the missing values of potential and kinetic energy.

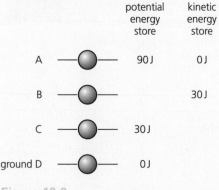

Figure 18.3

Answers online

Changes in energy

Kinetic energy

REVISED

The **kinetic energy** of a moving object can be calculated using the equation:

kinetic energy = 0.5 × mass × speed²

$$E_k = \frac{1}{2}mv^2$$

> kinetic energy, E_k, in joules, J
>
> mass, m, in kilograms, kg
>
> speed, v, in metres per second, m/s

Elastic potential energy

REVISED

The amount of **elastic potential energy** stored in a stretched spring can be calculated using the equation:

elastic potential energy = 0.5 × spring constant × extension²

$$E_e = \frac{1}{2}ke^2$$

(assuming the limit of proportionality has not been exceeded).

> elastic potential energy, E_e, in joules, J
>
> spring constant, k, in newtons per metre, N/m
>
> extension, e, in metres, m

Gravitational potential energy

REVISED

The **gravitational potential energy** gained by an object raised above the ground can be calculated using the equation:

gpe = mass × gravitational field strength × height

$$E_p = mgh$$

> gravitational potential energy, E_p, in joules, J
>
> gravitational field strength, g, in newtons per kilogram, N/kg
>
> height, h, in metres, m

Examples

1   Calculate the gravitational potential energy gained by a boy of mass 55 kg who climbs a flight of stairs 3.8 m high.
2   Calculate the elastic potential energy stored in a spring extended by 6.0 cm, which has a spring constant of 50 N/m.
3   Calculate the change in kinetic energy when a car of mass 1400 kg slows from 20 m/s to 10 m/s.

Answers

1   $E_p = mgh$

$$= 55 \times 9.8 \times 3.8$$

$$= 2048\,J \approx 2000\,J$$

2   $E_e = \frac{1}{2}ke^2$

$$= \frac{1}{2} \times 50 \times (0.06)^2$$

$$= 0.09\,J$$

3   $E_k = \frac{1}{2}mv_1^2 - \frac{1}{2}mv_2^2$

$$= \frac{1}{2} \times 1400 \times 20^2 - \frac{1}{2} \times 1400 \times 10^2$$

$$= (700 \times 400) - (700 \times 100)$$

$$= 210\,000\,J \text{ or } 210\,kJ$$

Typical mistake

You must always convert cm to m to get the energy in joules, J.

Exam tip

You should express your answer to the same number of significant figures as the data in the question. However, you will not be penalised over significant figures unless the question specifically asks for them.

Energy transfers and calculations

The equations for kinetic energy, elastic potential energy and gravitational potential energy may also be used to make calculations and predictions when a process causes energy to be transferred from one energy store to another.

Gravitational potential and kinetic energy stores

Example

A ball of mass 0.15 kg is thrown vertically upwards with a speed of 20 m/s.
1 Calculate the kinetic energy of the ball.
2 Calculate the maximum height to which the ball rises.

Answer

1 $E_k = \frac{1}{2}mv^2$

$\qquad \frac{1}{2} \times 0.15 \times 20^2$

$\qquad = 30\,J$

2 The energy in the ball's kinetic store is all transferred to the ball's gravitational potential energy store when it reaches its maximum height.
So the maximum E_p is 30 J.

$E_p = mgh$

$30 = 0.15 \times 9.8 \times h$

$h = \dfrac{30}{0.15 \times 9.8}$

$\qquad = 20.4\,m \approx 20\,m$

Maths note

The problem with the ball can be solved more quickly as follows:

E_k is transferred to E_p.
So,

$\frac{1}{2}mv^2 = mgh$

$h = \dfrac{v^2}{2g}$

$\quad = \dfrac{20^2}{2 \times 9.8}$

$\quad = 20.4\,m \approx 20\,m$

Now test yourself

4 Calculate the increase in the gravitational energy store of a girl of mass 45 kg who climbs to the top of the Shard, which has a height of 310 m. Gravitational field strength $g = 9.8\,N/kg$.
5 Calculate the kinetic energy of a bullet with a mass of 20 g, travelling at a speed of 700 m/s.
6 A car with a mass of 900 kg increases its speed from 12 m/s to 18 m/s. Calculate the increase in the car's kinetic energy store. Express your answer in kilojoules.
7 A car suspension spring has a spring constant of 1800 N/m. Calculate the elastic potential energy stored in the spring when it is compressed by 10 cm.
8 A stone of mass 0.08 kg is dropped from a bridge into a river 12 m below.
   (a) Calculate the stone's gravitational potential energy 12 m above the river.
   (b) Calculate the speed of the stone as it hits the river.
   (c) The stone comes to rest on the riverbed. Explain where the stone's original gravitational potential energy is stored now.

9 Figure 18.4 shows a gymnast bouncing on a trampoline. She has a mass of 52 kg. After dropping from a height of 4.0 m, the trampoline has been stretched by 0.9 m.

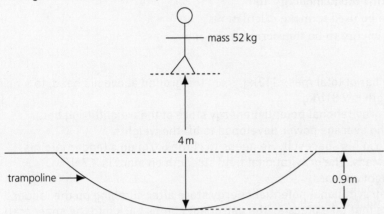

mass 52 kg

4 m

trampoline →

0.9 m

Figure 18.4

(a) Calculate the gymnast's potential energy, 4.0m above her lowest point.
(b) State the elastic potential energy stored in the trampoline after it has been stretched by 0.9 m.
(c) Use your answer to (b) to calculate the spring constant for the trampoline.

10 A bow has a spring constant of 300 N/m; it is used to shoot an arrow of mass 30 g. The bow string is pulled back 0.9 m before the arrow is released.
(a) Calculate the elastic potential energy stored in the bow when the string has been pulled back by 0.9 m.
(b) Assuming all the bow's elastic potential energy is transferred to the arrow's kinetic energy, calculate the arrow's speed when it is released.

Answers online

Work and power

Work

A force does work on an object when the force causes an object to move in the direction of the force.

work = force × distance moved in the direction of the force

$W = Fs$

When work is done by a force, the energy store of an object changes. For example:
- When 400 J of work is done to lift a box, the gravitational potential energy store of the box increases by 400 J.
- When 5000 J of work is done to accelerate a car, the kinetic energy store of the car increases by 5000 J.

work, W, in joules, J
force, F, in newtons, N
distance, s, in metres, m

Exam tip

Note, the equation for work will only be examined in the second paper. The equation is included here to help you with the definition of power.

Power

Power is the rate at which energy is transferred or the rate at which work is done.

$$power = \frac{energy\ transferred}{time}$$

or

$$power = \frac{work\ done}{time}$$

$$P = \frac{E}{t}$$

or

$$P = \frac{W}{t}$$

power, P, in watts, W
energy, E, transferred in joules, J
work done, W, in joules, J
time, t, in seconds, s

Now test yourself

11 State the correct unit for each of the following:
  (a) force
  (b) power
  (c) energy
  (d) work.
12 A weightlifter lifts a weightlifting bar of total mass 110 kg, from the ground above his head, to a height of 2.3 m. Gravitational field strength is 9.8 N/kg.
  (a) Calculate the increase in the gravitational potential energy store of the weightlifting bar.
  (b) The lift took 1.7 s. Calculate the average power developed to lift the weights.
13 An astronaut has a mass of 120 kg in his spacesuit. He needs to climb 8.0 m up a ladder into his spacecraft, which has landed on Mars. The gravitational field strength on Mars is 3.7 N/kg.
  (a) Calculate the astronaut's weight on Mars.
  (b) Calculate the increase to his gravitational potential energy store after climbing up the ladder.
  (c) Describe the energy transfers that take place as the astronaut climbs back into the spacecraft.
14 A crane lifts a weight of 15 000 N through a height of 28 m in 84 s. Calculate the output power of the crane in kW.
15 An electrical heater transfers 3000 J of thermal energy to a cup of water in 2 minutes. Calculate the power of the heater.

Answers online

Energy changes in systems

The amount of thermal energy stored in or released from a system as its temperature changes can be calculated using the equation:

change in thermal energy = mass × specific heat capacity × temperature change

$$\Delta E = mc\Delta\theta$$

change in thermal energy, ΔE, in joules, J

mass, m, in kilograms, kg

specific heat capacity, c, in joules per kilogram per degree Celsius, J/kg°C

temperature change, $\Delta\theta$, in degrees Celsius, °C

The specific heat capacity of a substance is the amount of energy required to raise the temperature of one kilogram of the substance by one degree Celsius.

Required practical 14

An investigation to measure the specific heat capacity of a material

There are several ways to measure specific heat capacity. All methods rely on the same principle: the decrease in one energy store (or work done) leads to the increase in thermal energy store of another material.

The apparatus shown in Figure 18.5 is used to warm up a steel block.

Exam tip

You should be able to describe an experiment and calculate specific heat capacity.

→

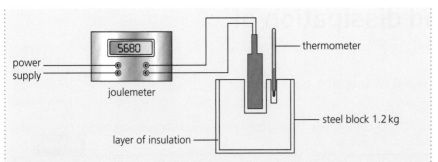

Figure 18.5

The following results were recorded:
- Electrical work done by the power supply to heat the steel was 5680 J.
- The initial temperature of the steel block was 18.2 °C.
- The final temperature of the steel block was 27.9 °C.
- Mass of the block was 1.2 kg.

$$\Delta E = mc\Delta\theta$$

$$\Delta\theta = 27.9 - 18.2 = 9.7\,°C$$

$$5680 = 1.2 \times c \times 9.7$$

$$c = \frac{5680}{1.2 \times 9.7}$$

$$= 490\,J/kg\,°C$$

The accepted value for the specific heat capacity of steel is 450 J/kg °C.

The measured value in a school laboratory is usually higher than the accepted value because:
- no account is made of thermal energy transfer to the surroundings
- the thermometer and heater require energy to warm them up too.

These are examples of systematic errors, because they always affect the answer by the same amount.

Now test yourself

TESTED

16 State the unit for specific heat capacity.
17 Use the data below to answer the following questions.
    Specific heat capacity of water is 4200 J/kg °C
    Specific heat capacity of air is 1000 J/kg °C
    (a) Calculate the energy required to warm up 90 kg of air in a room from 7 °C to 23 °C.
    (b) A kettle is filled with 0.8 kg of water at a temperature of 18 °C. The kettle has a power rating of 2200 W and it is switched on for 2 minutes.
        (i) Calculate the electrical work done by the kettle in two minutes.
        (ii) State the thermal energy transferred to the water; what assumptions are you making?
        (iii) Calculate the temperature of the water after 2 minutes of heating.
    (c) A heater transfers 12000 J of energy to a block of copper of mass 1 kg. During the heating the temperature of the copper rises from 22 °C to 52 °C. Calculate the specific heat capacity of copper.
18 Your coffee has cooled down to a temperature of 30 °C, and you like to drink it at a temperature of 50 °C. Your coffee has a mass of 0.2 kg and a specific heat capacity of 4000 J/kg °C.
    (a) Your microwave oven has a power rating of 800 W. Calculate how long you need to heat the coffee to reach a temperature of 50 °C.
    (b) After heating, you find that the temperature is only 47 °C. Explain why the temperature is lower than your calculation predicted.

Answers online

Conservation and dissipation of energy

Energy can be transferred usefully, stored or dissipated, but cannot be created or destroyed.

Storage and dissipation

You have already met the idea of energy being stored, for example in the kinetic energy of a moving car, or in the elastic potential energy of a stretched spring (page 223).

When energy is transferred from one store to another, energy can be dissipated or wasted. When energy is dissipated, it is stored in less useful ways.

> **Example**
>
> Chemical energy is stored in petrol; we want that energy to do work against resistive forces to keep a car moving. But when the petrol burns, energy is also transferred to the thermal energy store of the surroundings. That energy is wasted, as we did not want to produce it, and we cannot recapture it to do anything useful.

Energy saving

Wherever possible, we try to avoid the **dissipation** of energy, so that we maximise the useful transfer of energy from one store to another.

In the home

When we heat our home, eventually all the energy in the home's thermal store will be dissipated to the outside of the house. We get the best value for our heating bill, and we avoid wasting energy resources, if we slow down the process of thermal energy transfer.

- We insulate the loft.
- Carpets insulate the floors.
- Windows and doors are draught proofed.
- Thick walls reduce energy transfer.
- The cavity between the inside and outside wall of the house can be insulated.
- Double glazing reduces energy transfer through the windows, by trapping a layer of air between two panes of glass.

Insulation usually involved trapping a layer of air in fibres. Air carries energy efficiently by convection, but when air is trapped it is a good insulator, because air has a very low **thermal conductivity**.

> **Dissipation** is the wasting and spreading out of thermal energy into the surroundings.

> The **thermal conductivity** of a material is a measure of how quickly energy is transferred by conduction through it. Metals have very high conductivities. Brick and glass have lower conductivities, but energy still flows through them fast enough to cool a house down on a cold day.

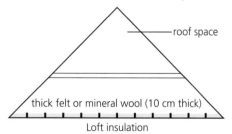

Figure 18.6 Loft insulation.

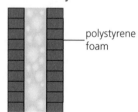

Figure 18.7 Polystyrene foam acts as an insulator in the cavity between the walls.

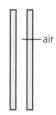

Figure 18.8 Air acts as an insulator in double-glazed windows.

Answers and quick quizzes at **www.hoddereducation.co.uk/myrevisionnotesdownloads**

Car design

We try to get the most useful energy out of petrol that we can.

- Cars are fuel efficient, which means less energy is dissipated to the surroundings.
- Cars are streamlined to reduce air resistance.
- Moving parts are lubricated with oil to reduce friction.

H

By reducing the dissipation of energy we increase the efficiency of the intended energy transfer. In the example of the car, streamlining and lubrication increase the amount of energy available to do work against resistive forces.

Efficiency

The energy efficiency for any energy transfer can be calculated using the equation:

$$efficiency = \frac{useful\ output\ energy\ transfer}{total\ input\ energy\ transfer}$$

Efficiency can also be calculated using the equation:

$$efficiency = \frac{useful\ power\ output}{total\ power\ input}$$

Example

A steam engine uses coal as its fuel. When the chemical store of the coal transfers 200 kJ of energy 24 kJ of work is done against resistive forces. Calculate the efficiency.

Answer

$$efficiency = \frac{useful\ output\ energy\ transfer}{total\ input\ energy\ transfer}$$

$$= \frac{24}{200}$$

$$= 0.12\ or\ 12\%$$

Exam tip

You can express efficiency as a decimal or as a percentage. Since efficiency is a ratio of energies (or powers) it has no unit.

Now test yourself

19 Explain the meaning of the phrase *energy dissipation*.
20 (a) List three ways in which we reduce energy losses from our homes.
    (b) Explain how each of your choices reduces energy dissipation.
21 Explain how a car can be designed to be more efficient in its use of petrol.
22 Define *efficiency*.
23 A car is supplied with 10 kg of fuel. A kilogram of fuel stores 4.5 MJ of chemical energy. The efficiency of the car is 30%.
    (a) Calculate the amount of energy available for useful work against resistive forces.
    (b) Calculate the amount of energy dissipated after all the fuel has been used.

→

24 The human body is about 25% efficient in transferring energy from the body's chemical store, to allow the body to do mechanical work.

(a) A boy of mass 60 kg climbs a tower of height 35 m. Calculate the energy in his gravitational potential store after the climb is completed. Gravitational field strength is 9.8 N/kg.

(b) Calculate the amount of chemical energy transferred for the boy's climb.

(c) Calculate the energy dissipated during the boy's climb.

(d) Assuming all the dissipated energy is transferred to the boy's thermal store, calculate the increase in his temperature. The specific heat capacity of a boy is 4000 J/kg °C.

Answers online

National and global energy resources

When we discuss energy resources, we are often interested in stores of energy that we can use to generate electrical power.

An energy store sets in motion a resource such as a gas or water that moves past a turbine which drives an electrical generator.

Non-renewable energy resources

REVISED

Much of our electricity in the UK comes from the fossil fuels, coal, oil and gas. These fuels store chemical energy. To release the energy, the fossil fuels must be burned. Once the fuels are burned, they are gone forever, because these fuels have taken millions of years to form. Nuclear fuels, uranium and plutonium, are also non-renewable, but there is a plentiful supply. These fuels are described as **non-renewable energy resources**, because there is a finite supply.

> **Non-renewable energy resources** will run out, because there are finite reserves, which cannot be replenished.
>
> **Renewable energy resources** will never run out, because these can be replenished.

Renewable energy resources

REVISED

By contrast **renewable energy resources** will never run out. We obtain renewable energy from the Sun, tides, waves, rivers and waterfalls, from the wind and from thermal energy in the Earth itself.

Using fuels

REVISED

The main uses of fuels are as follows:

● **Electricity**. In the UK electricity is generated using different energy resources: just over half is generated from fossil fuels, about 20% is generated from nuclear fuel and the rest is made using renewable resources.

● **Transport**. Fuels such as petrol, diesel and kerosene are produced from oil. These fuels drive our cars, trains and planes. Electricity is also used to run trains, and rechargeable batteries in electric cars are used.

● **Heating**. Most of our home heating is provided by gas and electricity. Some homes have oil-fired boilers, or burn solid fuels such as coal and wood.

> **Exam tip**
>
> Common nuclear fuels are uranium and plutonium.

Reliability of energy resources

REVISED

To generate electricity we need reliable energy resources. Fossil fuels are reliable, as we can mine coal and extract oil and gas from wells. However, our fuel resources might run out in the next hundred years, or become very expensive.

Tidal power is reliable, because we have high tides twice a day. However, the times of the high tides change each day, so the peak of electricity generation might not coincide with peak demand.

Solar, wind and hydroelectric power make useful contributions to electricity generation in many countries. But these are not reliable: there is less solar energy available in winter or on cloudy days; wind strength varies considerably; in some countries there is less hydroelectric power available in winter as rivers freeze.

> **Exam tip**
>
> Make sure you understand the advantages and disadvantages of resources used to generate electricity.

Environmental issues

REVISED

There is an increasing amount of evidence to show that the Earth is warming up (global warming). It seems likely that the Earth will soon be about 2 °C warmer (on average) than it was 50 years ago. Most scientists think that global warming is linked to the production of carbon dioxide and other **greenhouse gases** that trap radiation in the Earth's atmosphere.

As a result of global warming, many countries are committed to reducing the use of fossil fuels, and want to generate their electricity using renewable energy resources.

> A **greenhouse gas** is one that traps radiation in the Earth's atmosphere and, therefore, contributes to global warming. Carbon dioxide and methane are examples of greenhouse gases.

Renewable energy resources also have their impact on the environment.
- Wind turbines can be noisy and people object to them spoiling the look of the countryside.
- Tidal barrages that trap water at high tides can affect the habitat of wildlife.
- Hydroelectric dams affect the flow of rivers, and lakes made behind dams have flooded small towns, causing communities to be relocated.

The production of electricity requires a balance between our needs and environmental issues.

Political and economic issues

REVISED

We can only solve the problems caused by the production of greenhouse gases if all countries agree. The UK plans to stop producing electricity from coal by 2025. However, unless all countries do the same, there will still be problems with global warming.

One reason countries continue to burn coal is cost. Coal-fired power stations are cheaper to run than nuclear ones, or renewable energy resources. To solve environmental issues, we need to be prepared to pay more for our electricity.

Now test yourself

TESTED

25(a) Explain what a renewable energy resource is. Give an example.
  (b) Explain what a non-renewable energy resource is. Give an example.
26(a) State two advantages of using a coal-fired power station.
  (b) State two environmental problems associated with coal-fired power stations.

Answers online

Summary

- Energy is an idea that cannot be described by a single process. However, we pay an enormous amount of attention to energy because it is conserved.
- There are different stores of energy. In any process, energy can be transferred from one store to another, but energy is never created or destroyed.
- Energy stores include: kinetic, chemical, internal (or thermal), gravitational potential, magnetic, electrostatic, elastic potential and nuclear.
- Energy can be transferred from one store to another by: mechanical work, electrical work, heating and radiation (mechanical and electromagnetic).
- The amount of energy transferred to or from a store may be calculated from these equations:

  kinetic energy $E_k = \frac{1}{2}mv^2$

  gravitational potential energy: $E_p = mgh$

  elastic potential energy: $E_e = \frac{1}{2}ke^2$

  thermal energy: $\Delta E = mc\Delta\theta$
- Mechanical work done is calculated using: $W = Fs$
- Power is calculated using the equations:

  $P = \dfrac{\text{energy}}{\text{time}}$

or

$P = \dfrac{\text{work}}{\text{time}}$

- Units:
  - power – watts (W)
  - energy – joules (J)
  - work – joules (J)
  - force – newtons (N)
- In many processes there are unwanted energy losses in the form of the transfer of thermal energy. This energy cannot be recovered into a useful form.

  $\text{efficiency} = \dfrac{\text{useful output energy transfer}}{\text{useful input energy transfer}}$

or

$\text{efficiency} = \dfrac{\text{useful power output}}{\text{total power input}}$

- Energy resources can be non-renewable – oil, gas, coal and nuclear fuels.
  Or, energy resources can be renewable – wind, waves, hydroelectric, geothermal, solar, for example.
- Fuels are used for transport, heating and generating electricity.

Exam practice

1 Which of the following is required for a hydroelectric power station?
   A Sunlight
   B A supply of falling water
   C A supply of hot water from the Earth [1]
2 An electric car uses a battery to power it. The car accelerates from rest.
   (a) Choose words from the list below that describe the energy transfers while the car is accelerating. [3]

kinetic energy   gravitational   potential   thermal   chemical   elastic potential

   The battery has a store of .... energy. When the car accelerates, the motor transfers some useful energy to increase the car's store of .... energy. Some energy is wasted which increases the .... energy store of the motor and is surroundings.

   (b) When the car is accelerating the motor's output power is 4800 W. The motor transfers energy into useful energy at a rate of 1200 W.
      Calculate the efficiency of the car's motor. [2]
3 (a) State one advantage and one disadvantage of using nuclear power. [2]
   (b) A nuclear power station has a power output of 3000 MW. A wind turbine has a maximum power output of 2 MW.
      (i) How many watts, W, are there in a megawatt, MW? [1]
      (ii) How many wind turbines, working at their maximum rate, produce the same power as a nuclear power station? [2]
      (iii) Explain one advantage and one disadvantage of using wind turbines instead of a nuclear power station. [2]

4 (a) (i) Name one renewable fuel and one non-renewable fuel. [2]
  (ii) Explain what the words *renewable* and *non-renewable* mean in this context. [2]
  (b) Explain two reasons why governments may wish to increase the amount of electricity generated using renewable energy resources. [2]

5 When we heat our homes, often energy is wasted.
Choose three ways in which energy can be wasted, and explain how that waste can be reduced. [6]

6 A heater is used to increase the temperature of a block of tin. The graph in Figure 1.11 shows how the temperature of the tin rises, with the energy transferred by heating. The energy is measured by the joulemeter.

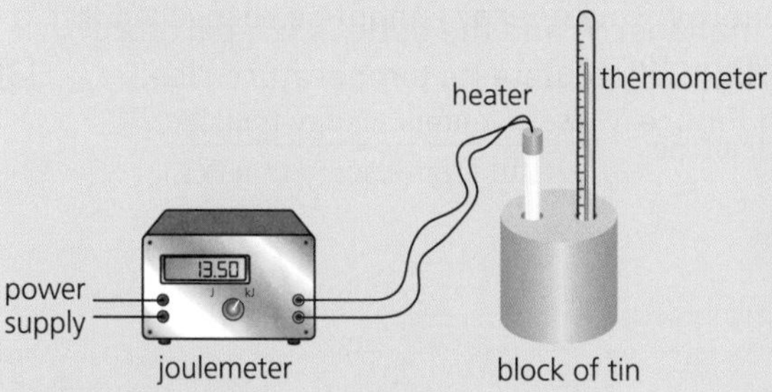

Figure 18.9

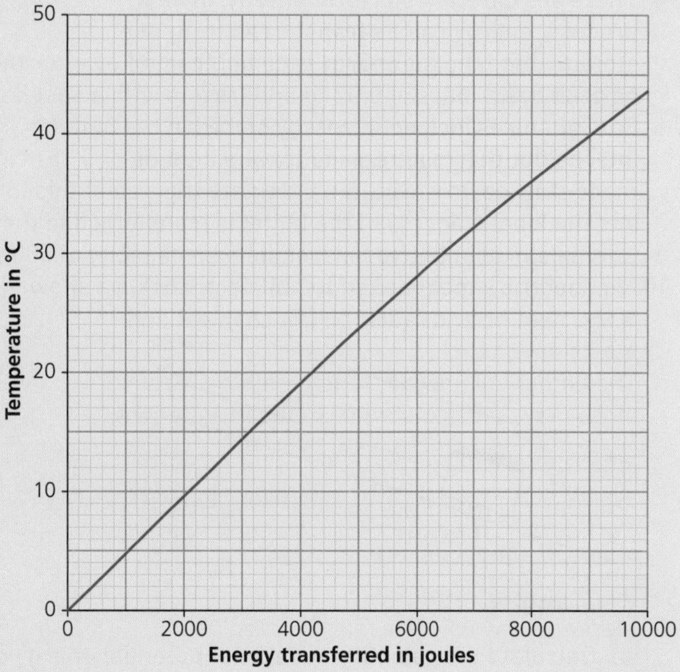

Figure 18.10

(a) Use the graph to calculate the temperature rise after 2000 J of energy has been transferred by the heater. [1]
(b) The block of tin has a mass of 0.8 kg. Calculate its specific heat capacity. [3]
(c) The graph is not linear. Give a reason for the graph being non-linear.
(d) Suggest a possible improvement to this experiment. [1]

7 A commercial hovercraft runs between Portsmouth and the Isle of Wight. The hovercraft has a mass of 60 000 kg.

Figure 18.11

(a) A hovercraft approaches the beach with its engines off. It comes to a halt after it has gone up a height of 5.0 m.
Calculate the increase of gravitational potential energy stored by the hovercraft after it has gone up 5.0 m. Gravitational field strength is 9.8 N/kg. [3]
(b) Assuming that the kinetic energy store of the hovercraft has been converted to a store of gravitational potential energy at the top of the beach, calculate the hovercraft's speed just as it approached the beach. [3]

8 A spring has a spring constant of 60 N/m; it has an unstretched length of 0.2 m.
(a) Show that the spring stores elastic potential energy of 0.3 J when it has been stretched to a length of 0.3 m. [3]
(b) The spring is now attached to a trolley as shown in Figure 18.12.

→

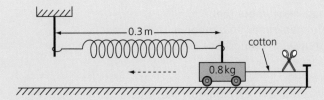

Figure 18.12

Calculate the maximum speed of the trolley after the string is cut. [3]

9 A bullet has a mass of 0.02 kg and travels with a speed of 400 m/s.
   (a) Calculate the kinetic energy stored in the bullet. [3]
   (b) The bullet hits a tree and travels a depth of 0.25 m into the tree. A resistive force does work to slow down the bullet. Calculate the size of this force. [3]
   (c) The kinetic energy of the bullet is transferred to thermal energy and the bullet's temperature rises. The bullet has a specific heat capacity of 500 J/kg °C. Calculate its temperature rise. [3]

10 An electric winch is used to pull up a truck, as shown in Figure 1.14.

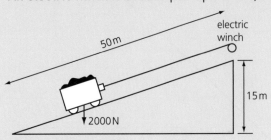

Figure 18.13

   (a) Calculate the gain in gravitational potential energy of the truck after it has been pulled up 15 m. [3]
   (b) The winch uses a 5 kW electric supply and takes 12 s to pull the truck 50 m along the slope. Calculate the electrical work done by the winch. [3]
   (c) Calculate the efficiency of the winch. [2]

11 A power supply is used to heat 0.1 kg of water in an insulated beaker. The water has a temperature of 20 °C.
   Use the information in the diagram to calculate how long it takes for the water to warm to 50 °C. Water has a specific heat capacity of 4200 J/kg °C. [6]

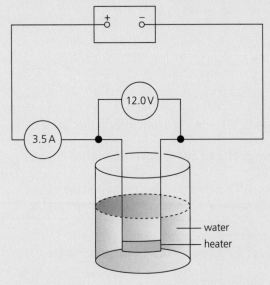

Figure 18.14

Answers and quick quiz 18 online

ONLINE

19 Electricity

Electrical power is an integral part of our lives. It fills our world with artificial light and information, and allows us to be entertained at any time of the day.

Current, potential difference and resistance

Figure 19.1 shows the standard **circuit symbols** you need to know.

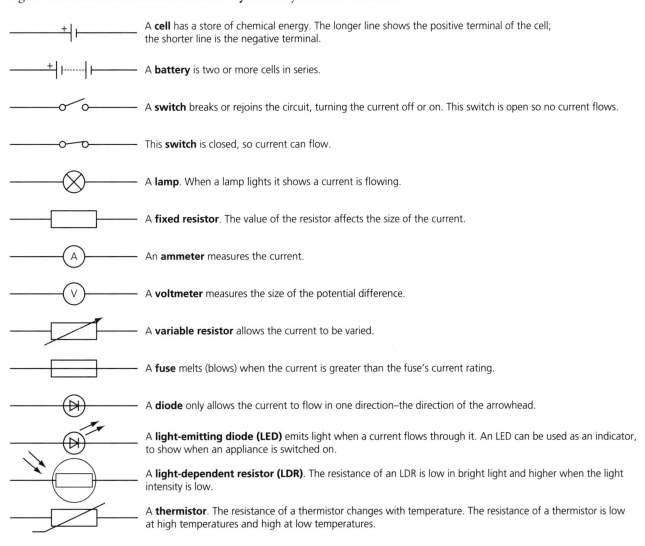

A **cell** has a store of chemical energy. The longer line shows the positive terminal of the cell; the shorter line is the negative terminal.

A **battery** is two or more cells in series.

A **switch** breaks or rejoins the circuit, turning the current off or on. This switch is open so no current flows.

This **switch** is closed, so current can flow.

A **lamp**. When a lamp lights it shows a current is flowing.

A **fixed resistor**. The value of the resistor affects the size of the current.

An **ammeter** measures the current.

A **voltmeter** measures the size of the potential difference.

A **variable resistor** allows the current to be varied.

A **fuse** melts (blows) when the current is greater than the fuse's current rating.

A **diode** only allows the current to flow in one direction–the direction of the arrowhead.

A **light-emitting diode (LED)** emits light when a current flows through it. An LED can be used as an indicator, to show when an appliance is switched on.

A **light-dependent resistor (LDR)**. The resistance of an LDR is low in bright light and higher when the light intensity is low.

A **thermistor**. The resistance of a thermistor changes with temperature. The resistance of a thermistor is low at high temperatures and high at low temperatures.

Figure 19.1

Electrical charge and current

Figure 19.2 shows a simple circuit.

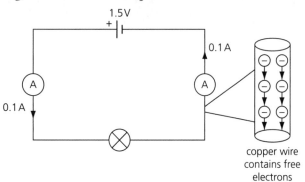

Figure 19.2

In this circuit a cell provides a potential difference (p.d.) of 1.5 V to drive a current of 0.1 A.

- The potential difference is a measure of the electrical work done by the cell to drive the current round the circuit.
- The current is a measure of the rate at which charge flows round the circuit.
- The charge is measured in coulombs.
- We have a convention that current flows from positive to the negative terminal of the cell. But when electrons flow, they travel from the negative to the positive terminal of the cell.

Charge, current and time are linked by this equation:

charge flow = current × time

$$Q = It$$

charge flow, Q, in coulombs, C

current, I, in amperes, A (amp is acceptable for ampere)

time, t, in seconds, s

Small currents are measured in milliamps (mA)

Now test yourself

1 Draw a circuit diagram to show a cell, an ammeter, a lamp and a resistance connected in series.
2 In an electrical circuit, a charge of 12 C flows round a circuit in 2 minutes.
   Calculate the current in
   (a) amps
   (b) milliamps.

Answers online

Current, resistance and potential difference

The current, I, through a component depends both on the resistance, R, of the component and the potential difference, V, across the component. The greater the resistance of the component, the smaller the current, for a particular potential difference.

The current, potential difference and resistance are linked by the equation:

potential difference = current × resistance

$$V = IR$$

Large resistances may be measured in kilohms (kΩ) and megohms (MΩ).
- $1\,k\Omega = 1000\,\Omega$
- $1\,M\Omega = 1\,000\,000\,\Omega$

potential difference, V, in volts

current, I, in amperes (or amps)

resistance, R, in ohms, Ω

Ammeters and voltmeters

Figure 19.3 shows how an ammeter and voltmeter are connected to measure resistance. The ammeter must be in series with the resistor and the voltmeter must be in parallel with the resistor.

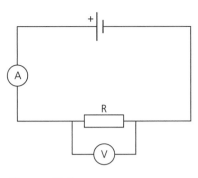

Figure 19.3

Example

In Figure 19.3 the voltmeter reads 12V and the ammeter reads 0.06A. Calculate the resistance.

Answer

$V = IR$

$12 = 0.06 \times R$

$R = \dfrac{12}{0.06}$

$= 200\,\Omega$

Required practical 14

You should have set up a circuit to investigate how the resistance of a given wire depends on its length. Figure 19.4 shows you the circuit to use.

You will have found that **the resistance of the wire is proportional to its length**.

For example: if a wire of length 40 cm has a resistance of 7.5 Ω, a length of 80 cm of the same wire has a resistance of 15.0 Ω.

You are also expected to be able to set up circuits to investigate combinations of resistors in series and in parallel. This is covered in Exam practice question 2 on page 196.

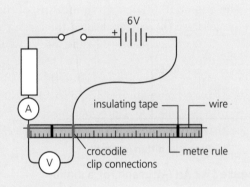

Figure 19.4

Resistors

REVISED

Ohmic conductors and non-ohmic conductors

The current through an ohmic conductor (at a constant temperature) is directly proportional to the potential difference across it. A graph of current against potential difference is a straight line (Figure 19.5). The resistance stays the same as the current changes.

The resistances of components such as **lamps**, **diodes**, **thermistors** and **light-dependent resistors** are not constant; the resistance changes with the current through the component. These are **non-ohmic conductors**.

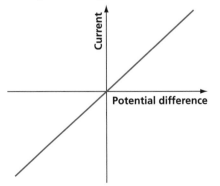

Figure 19.5 The *I–V* graph for an ohmic conductor at a constant temperature.

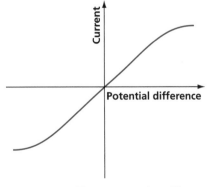

Figure 19.6 The current in a filament lamp is not proportional to the potential difference.

A filament lamp

When a current flows through a filament lamp, the filament heats up.
The resistance of a filament lamp increases as the temperature increases
(Figure 19.6).

A diode

A diode is a component that allows current to flow only one way (Figure 19.7).
A diode has a very high resistance in the reverse direction.

Light-dependent resistor (LDR)

The resistance of an LDR decreases as the light intensity increases
(Figure 19.8).

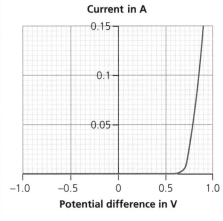

Figure 19.7 An *I–V* graph for a diode.

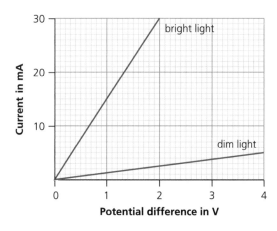

Figure 19.8 *I–V* graphs for an LDR in bright and dim light.

Thermistor

The resistance of a thermistor decreases as the temperature rises (Figure 19.9).

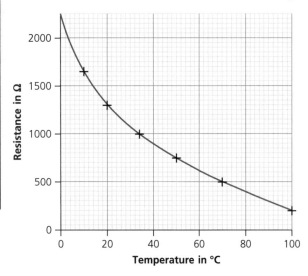

Figure 19.9 The resistance of a thermistor changes with the
temperature.

Answers and quick quizzes at **www.hoddereducation.co.uk/myrevisionnotesdownloads**

Required practical 16

You should be able to describe an experimental set-up and procedure that enables you to investigate the *I–V* characteristic graphs for a filament lamp, a diode and a resistance at a constant temperature.

This can be done using the apparatus shown in Figure 19.10.
- The current and potential difference values are recorded by reading the ammeter and voltmeter.
- The current can be altered by changing the number of cells or by changing the variable resistor.
- Using your data, you then plot an *I–V* graph for each component.

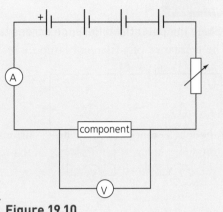

Figure 19.10

Now test yourself

3 (a) Explain what is meant by an *ohmic resistor*.
   (b) With reference to Figure 19.6, explain why a filament lamp is a non-ohmic resistor.
4 This question refers to a diode with the *I–V* characteristics shown in Figure 19.7.
   (a) Use the graph to calculate the potential difference across the diode, when a current of 0.1 A flows through it.
   (b) Calculate the resistance of the diode when 0.1 A is flowing through it.
5 In the list below there are five units for different electrical quantities.
   (a) Which is the correct unit for resistance?
   (b) Which is the correct unit for electrical charge?

   volt   coulomb   amp   ohm   watt

6 Describe an experiment to determine the *I–V* characteristics for a filament lamp. In your explanation you should state what apparatus you will use, and how you will use it. You should also explain what measurements you will take.

Answers online

Series and parallel circuits

There are two ways of connecting electrical components in a circuit, in **series** and in **parallel**.

Series circuits

For components connected in series:
- there is the same current through each component
- the potential difference of the power supply is shared between the components. If there are just two components then:

$$V_{supply} = V_1 + V_2$$

- the total resistance of two components is the sum of the resistance of each component:

$$R_{total} = R_1 + R_2$$

Example

Calculate the total resistance between A and B.

Figure 19.11

Answer

$$= 5 + 10 = 15\,\Omega$$

Example

State the potential difference across lamp 2 in Figure 19.12.

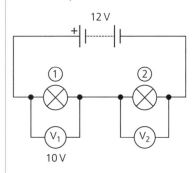

Figure 19.12

Answer

$$V_{supply} = V_1 + V_2$$

$$12 = 10 + V_2$$

$$V_2 = 12 - 10 = 2\,V$$

Parallel circuits

For components connected in parallel:
- the potential difference across each component is the same
- the total current through the whole circuit is the sum of the currents through the separate components
- the total resistance of two resistors in parallel is less than the resistance of the smaller individual resistor.

> **Typical mistake**
>
> Often students think that when two resistors are in parallel, the total resistance is the sum or the mean of the two. In fact, the total resistance is less than the smaller resistor.

Example

In Figure 19.13, state the value of the current going through lamp 2, and the potential difference across each lamp.

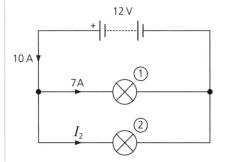

Figure 19.13

Answer

Current:

$$I_2 = 10 - 7 = 3\,A$$

Each lamp has 12V across it.

> **Typical mistake**
>
> In figure 19.13 students sometimes think the potential difference across each lamp is 6V. It is not: the potential difference across each is the 12V of the battery.

Now test yourself

7 Calculate the resistance between:
   (a) AB
   (b) CD.

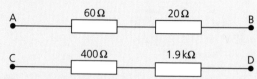
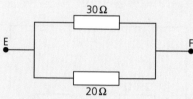

Figure 19.14

8 Which of the following correctly states the resistance between points E and F?

Figure 19.15

50 Ω   25 Ω   less than 20 Ω   between 30 Ω and 50 Ω

9 (a) Calculate the potential difference V_2 in Figure 19.16.
   (b) Calculate the two ammeter readings A_1 and A_2 in Figure 19.17.

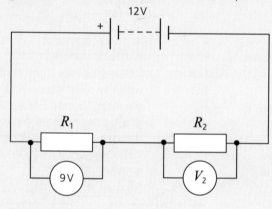
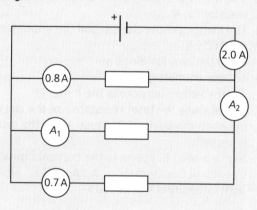

Figure 19.16   Figure 19.17

Answers online

Circuit calculations

You will be expected to use the rules about series and parallel circuits (above) to solve circuit problems. Some examples are given below.

Example

Calculate the readings on:
1 the ammeter, A
2 the voltmeter, V.

Answer

1 $V = IR$
  $4 = I \times 6$
  $I = \dfrac{4}{6}$
  $= 0.67\,A$

Figure 19.18

Typical mistake

Sometimes students combine the wrong p.d. with the wrong resistor. Remember, when you use $I = \dfrac{V}{R}$ to calculate the current, V is the p.d. across the resistor R: in Figure 19.18, 4V across the 6 Ω.

2 The p.d. across the 12 Ω resistor is:

$V = IR$

$= 0.67 \times 12$

$= 8\,V$

So the battery p.d., as measured by the voltmeter V, is: 8 + 4 = 12 V

Now test yourself

10 (a) A thermistor is connected into the circuit shown in
       Figure 19.19, when the temperature is 15 °C.
       Use the information in the diagram to calculate
       (i) the potential difference measured by the voltmeter, V
       (ii) the resistance of the thermistor.
    (b) The next day the temperature goes up to 25 °C. Explain
       what happens to the voltmeter reading.
11 In Figure 19.20, the switch S is left open.
    (a) State the currents measured by the ammeters,
        A_1 and A_2.
    (b) Use the information in the diagram to calculate the
        resistance, R.
    (c) The switch is now closed, and the ammeter A_1
        reads 0.1 A.
        State the new readings on:
        (i) the ammeter, A_2
        (ii) the voltmeter across the battery.
    (d) Explain why the total resistance of the circuit
        between the points AB is less when the switch
        is closed.
    (e) Explain what happens to the currents measured
        by each of the ammeters, A_1, A_2 and A_3, when the
        light intensity is increased.

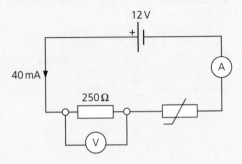

Figure 19.19

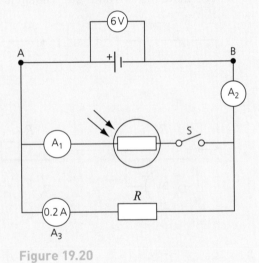

Figure 19.20

Answers online

Domestic use and safety

Direct and alternating potential difference

Power supplies can provide direct or alternating potential differences.
This is illustrated in Figure 19.21.

- The blue line shows a direct potential difference of 6 V. This will make
  a direct current (d.c.) flow in one direction through a resistor.
- The red line shows an alternating potential difference of 6 V. This
  changes direction so an **alternating current** (a.c.) flows first one way,
  then the other through a resistor. The peak value of the 6 V a.c. supply
  rises above 6 V to make up for the time when the potential difference is
  close to zero.

> **Alternating current (a.c.)** is
> current that flows one way
> and then the other.

Mains supply

The mains supply in the United Kingdom has a frequency of 50 Hz and a potential difference of about 230 V. A frequency of 50 Hz means that one cycle – as shown by the red curve in Figure 19.21, takes one-fiftieth of a second.

Mains electricity

REVISED

Most electrical appliances are connected to the mains using a three-core cable. Each of the wires inside the cable is colour coded for easy identification.
- Live wire – brown
- Neutral wire – blue
- Earth wire – green and yellow stripes

The live wire carries the alternating potential difference from the mains supply. The neutral wire completes the circuit. So the live and neutral wires carry the current to and from an electrical appliance.

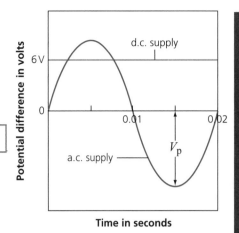

Time in seconds

Figure 19.21 V_p is the peak voltage.

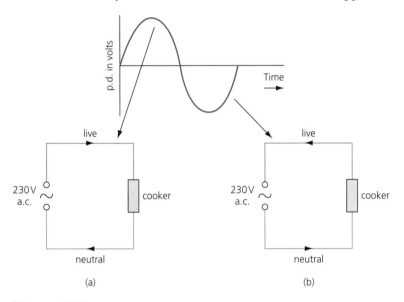

Figure 19.23

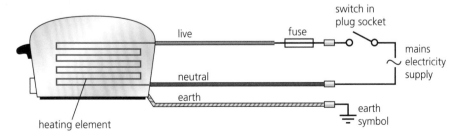

Figure 19.22 A cable has three wires; earth (green/yellow), neutral (blue), live (brown).

- The potential difference between the live wire and earth (0 V) is about 230 V. Even though an appliance is off and there is no current in the mains circuit, a live wire is dangerous. If you touch a live wire, current passes through you to earth, giving you a painful shock.
- The neutral wire is close to earth potential (0 V).
- The earth wire is at 0 V, and only carries a current if there is a fault.

Earthing

Any electrical appliance that has a metal case should be earthed. The toaster in Figure 19.24 has the earth wire connected to its metal case.

Figure 19.24

Any contact between the live wire and earth is potentially dangerous, because a large current passes to earth, which could start a fire.

Energy transfers

Power

The electrical power transferred by any electrical device is equal to the energy transferred per second.

The power transferred depends on the potential difference and current:

$$P = VI$$

power = (current)^2 × resistance

$$P = I^2R$$

> power, P, in watts, W
>
> potential difference, V, in volts, V
>
> current, I, in amperes, A (or amps)
>
> resistance, R, in ohms, Ω

Energy

When charge flows round a circuit, electrical work is done.

The energy transferred by electrical work depends on how long the appliance is switched on.

energy transferred = power × time

$$E = Pt$$

energy transferred = charge flow × potential difference

$$E = QV$$

> energy transferred, E, in joules, J
>
> time, t, in seconds, s
>
> charge flow, Q, in coulombs, C
>
> potential difference, V, in volts V

Examples

1 The information plate on a kettle is marked as follows:

    230V 50Hz   2650W

  Calculate the current drawn from the supply.
2 Calculate the energy transferred by the kettle (in Example 1) in 2 minutes.
3 When 150C of charge passes through a battery, 900J of energy is transferred. Calculate the potential difference of the battery.

Answers

1   $P = VI$

   $2650 = 230I$

   $I = \dfrac{2650}{230}$

   $= 11.5\,A$

2 $E = Pt$

   $= 2650 \times 120$

   $= 318\,000\,J$

   $= 318\,kJ \approx 320\,kJ$

3 $E = QV$

   $900 = 150 \times V$

   $V = 6\,V$

Typical mistake

Often power and energy are confused.

$$power = \frac{energy}{time}$$

Power is measured in joules per second or watts.

Energy is measured in joules.

Typical mistake

When using time in an equation, remember to turn the time into **seconds**.

Now test yourself

12 Which of the following is the correct unit for:
   (a) power   (b) energy   (c) charge?

   volt   amp   joule   ohm   watt   coulomb

13 Which of the following is an equivalent unit for:
   (a) volt   (b) amp?

joule × coulomb   coulomb/second   joule/coulomb   joule/second   coulomb × second

14 Calculate the power rating in watts of:
   (a) A fire that draws 8 A from a 230 V supply.
   (b) A lamp that draws 5 A from a 12 V supply.
15 An electric shower runs from a 230 V 15 A supply. Calculate the energy transferred to heat the water when someone has a shower for 3 minutes.
16 Calculate the electrical work done when:
   (a) 150 C of charge flows through a lamp with a potential difference of 12 V across it
   (b) a current of 8 mA flows through a 6 V battery charger for 4 hours.
17 A current of 0.1 A flows through a resistor of 220 Ω for 20 minutes. Calculate the energy transferred by the resistor.

Answers online

The National Grid

The National Grid is a system of cables and transformers that links power stations to consumers.

Electrical power is transferred through the National Grid.

Transformers step up the potential difference from the power station to the transmission cables (seen on overhead power lines). The high potential difference makes the current much lower; therefore less energy is wasted when the current is carried over long distances. Transformers in towns step down the high potential difference to the safe 230 V we use in our homes.

Summary

- You should know the circuit symbols shown on page 185.
- Current is a flow of charge.

  charge = current × time

  $$Q = It$$

- Current is measured in amps, A.
- Charge is measured in coulombs, C.
- Potential difference is measured in volts, V.

  potential difference = current × resistance

  $$V = IR$$

- Resistance is measured in ohms, Ω.
- When two components are in series:
  - the current is the same through each component

  - the potential difference of the power supply is shared between each component
  - the total resistance is the sum of the two resistances,

  $$R_{total} = R_1 + R_2$$

- When two components are connected in parallel:
  - the potential difference across each component is the same
  - the total current through the circuit is the sum of the currents through each component,

  $$I_{total} = I_1 + I_2$$

  - the total resistance is less than the resistance of the smallest individual resistor.

● A direct current flows one way round a circuit. An alternating current switches from one direction to the other.
● power = potential difference × current
$$P = VI$$
$$P = I^2R$$

● energy transferred = power × time
$$E = Pt$$
● energy transferred = charge flow × potential difference
$$E = QV$$

Exam practice

1 Figure 19.25 shows an electrical circuit.

(a) The two cells are identical. State the potential difference of one cell. [1]
(b) State the reading on the voltmeter. [1]
(c) State the reading on the ammeter. [1]
(d) Show by calculation that the resistance of the lamp is about 12 Ω. [3]
(e) Which of the following best describes the total resistance of the circuit? [1]

27 Ω   between 15 Ω and 12 Ω   less than 12 Ω

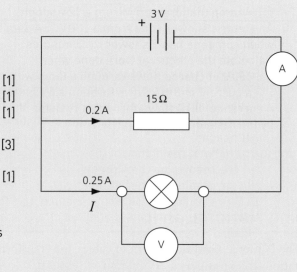

Figure 19.25

2 A student designs an experiment to investigate the effect of adding two resistors together in series. His circuits are shown in Figure 19.26.

(a) Use the information in Figures 19.26 (a) and 19.26 (b) to show that the values of the resistors R_A and R_B are:
(i) $R_A = 20\,\Omega$
(ii) $R_B = 40\,\Omega$. [3]

(b) The resistors are now put in series as shown in Figure 19.26 (c). The student predicts that the current in this circuit will be 0.10 A. Show by calculation how the student reached this hypothesis. [2]

(c) The student puts the two resistors in parallel as shown in Figure 19.26 (d). He now predicts that the total resistance of the circuit will be less than 20 Ω. Use the information in Figures 19.26 (a) and (b) to explain how the student reached his hypothesis. [3]

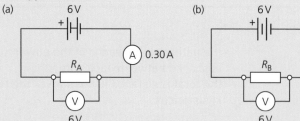

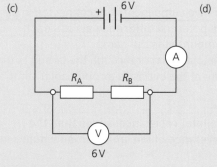

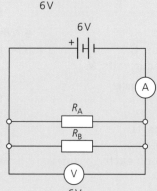

Figure 19.26

3 In Figure 19.27 the 9V battery supplies a direct current of 0.05 A to the circuit.

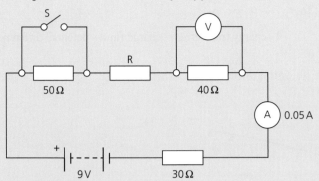

Figure 19.27

(a) Explain what is meant by *direct p.d.* [1]
(b) Use the information in the diagram to calculate the total resistance of the circuit. [3]
(c) Use your answer to part (b) to calculate the resistance, R. [1]
(d) The switch S is now closed. Explain what happens to each of the following (a calculation is not needed):
   (i) the total resistance of the circuit [1]
   (ii) the reading on the ammeter [1]
   (iii) the reading on the voltmeter. [1]

4 (a) Draw a circuit that you would use to obtain the data needed to draw a current–potential difference graph for a lamp. [3]
(b) Sketch *I–V* graphs for:
   (i) a filament lamp [1]
   (ii) a resistor at a constant temperature [1]
   (iii) a diode. [1]

5 Figure 19.28 shows three resistors connected across a 12 V cell.

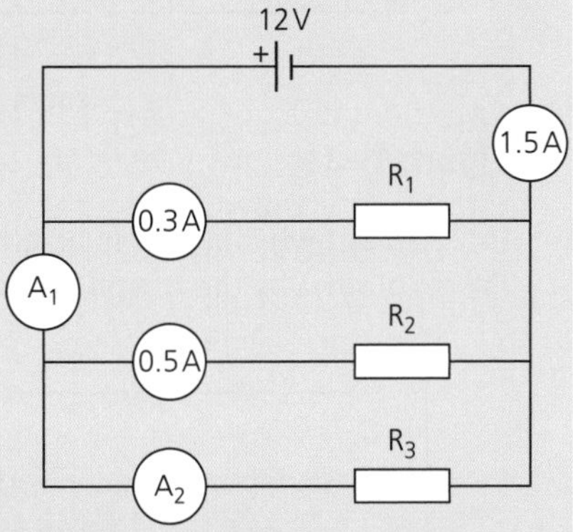

(a) Calculate the currents through the ammeters, A_1 and A_2. [2]
(b) Which resistance is greater, R_1 or R_2? Explain your answer. [2]
(c) (i) Calculate the resistance, R_2. [3]
   (ii) Calculate the power dissipated in the resistance, R_2. [3]

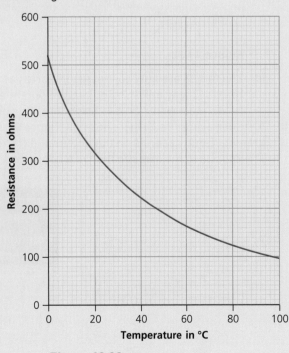

Figure 19.28

6 Figure 19.29 shows a circuit which includes a fixed resistor of 240 Ω and a component X. The resistance of X changes with temperature, as shown in Figure 19.30.

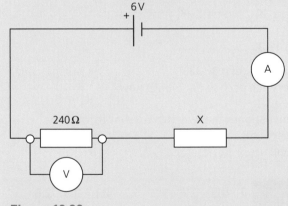

Figure 19.29

(a) The current through the ammeter is 0.01 A. Calculate the potential difference measured by the voltmeter, V. [2]
(b) Now calculate the potential difference across X. [1]
(c) Use your answer to (b) to show that X has a resistance of 360 Ω. [2]

Figure 19.30

(d) Use Figure 19.30 to calculate the temperature of X. [1]

(e) X is now warmed up. Explain what happens to:

(i) the total resistance of the circuit [1]

(ii) the ammeter reading, A [1]

(iii) the voltmeter, reading, V. [1]

7 Figure 19.31 shows how the current varies with potential difference for two different lamps, lamp X and lamp Y.

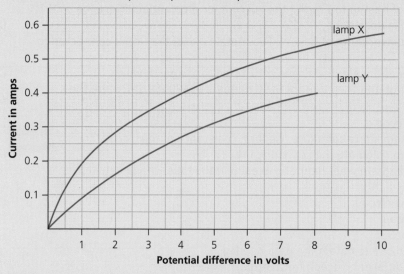

Figure 19.31

(a) Explain why lamp X is brighter than lamp Y in the circuit shown in Figure 19.32. [2]

(b) Explain why the graph shows that lamp Y has a higher resistance than lamp X. [2]

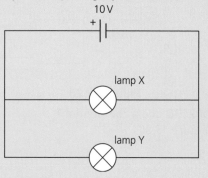

Figure 19.32

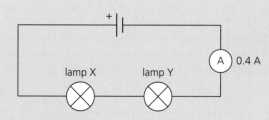

Figure 19.33

(c) The lamps are now connected in series as shown in Figure 19.33.

(i) Use the graph to calculate the potential difference across each bulb now. [2]

(ii) Explain which bulb is brighter in this circuit. [2]

8 A student uses the circuit shown in Figure 19.34 to investigate the way the current through a filament lamp depends on the potential difference across it.

The results of the investigation are shown in the table.

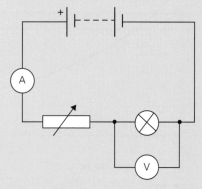

Current in amps	0	0.5	0.7	1.2	1.7	2.1	2.3	2.8	3.1
Potential difference in volts	0	0.3	0.8	1.8	3.7	5.0	6.5	9.0	11.0

Figure 19.34

(a) Which is the dependent variable, current or potential difference? [1]

(b) Plot a graph of current (*y*-axis) against potential difference (*x*-axis). Draw a line of best fit through the points. [4]

(c) (i) The student made an error in one ammeter measurement. State what the correct reading should have been.

(ii) Name the type of error that has occurred here and explain what action can be taken to reduce the likelihood of such errors. [1]

(d) A student extends the line of best fit to predict the current for a potential difference of 12 V. Explain why this prediction is unreliable. [2]

9 An electric fire element has a power input of 2.3 kW. The metal case of the fire is connected to the earth wire.

(a) Explain why the metal case of the electric fire is connected to the earth wire. [2]

(b) The charge that flows through the fire element in 10 minutes is 6000 C. Calculate the resistance of the element. [5]

Answers and quick quiz 19 online

ONLINE

20 Particle model of matter

Changes of state and the particle model

Density

The density of a material is defined by the equation:

$$\text{density} = \frac{\text{mass}}{\text{volume}}$$

$$\rho = \frac{m}{V}$$

> density, ρ, in kilograms per metre cubed, kg/m^3
>
> mass, m, in kilograms, kg
>
> volume, V, in metres cubed, m^3

Example

Mercury has a density of 13 600 kg/m^3. Calculate the mass of 0.002 m^3 of mercury.

Answer

$$\rho = \frac{m}{V}$$

So

$$13\,600 = \frac{m}{0.002}$$

$$m = 13\,600 \times 0.002$$

$$= 27.2\,\text{kg}$$

Solids, liquids and gases

- **Solid.** In a solid, atoms (or molecules) are packed close together in a regular structure. The atoms cannot move from their fixed positions, but they can vibrate. The atoms are held together by strong forces, so it is difficult to change the shape of a solid.
- **Liquid.** The atoms (or molecules) in a liquid are close together. Forces keep the atoms in contact, but the atoms are free to move. A liquid can flow and change shape to fit into any container. Because the atoms are close together, it is difficult to compress a liquid.
- **Gas.** In a gas the atoms (or molecules) are separated by relatively large distances. The forces between the atoms are small. The atoms are in a constant state of random motion. A gas can expand to fill any volume, and a gas is easy to compress.

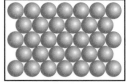

Solid

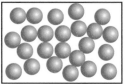

Liquid

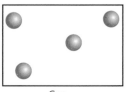

Gas

Figure 20.1

The density of a material can be explained by the particle model of matter:
- Gases have low densities because atoms and molecules are far apart in the gaseous state.
- Metals such as gold are very dense because:
  - the atoms are packed close together
  - each atom has a high mass.

Required practical 17

You should be able to describe how to take appropriate measurements and then calculate the densities of liquids and solids.

The density of a liquid

The cylinder in Figure 20.2 has a mass of 152.6 g when empty and a mass of 169.2 g when it has 20 cm³ of liquid in it. Calculate the density of the liquid in g/cm³.

mass of the liquid = 169.2 − 152.6

$$= 16.6 \text{ g}$$

$$\rho = \frac{m}{V}$$

$$= \frac{16.6}{20}$$

$$= 0.83 \text{ g/cm}^3$$

(a)

(b)

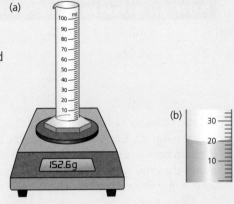

Figure 20.2

The density of a regular solid

A metal block has a height of 5.1 cm, a length of 10.7 cm and width of 9.3 cm. The mass of the block is 3.83 kg. Calculate the density of the metal in kg/m³.

volume of block = 0.051 × 0.107 × 0.093

$$= 5.1 \times 10^{-4} \text{ m}^3$$

$$\rho = \frac{m}{V}$$

$$= \frac{3.83}{5.1 \times 10^{-4} \text{ m}^3}$$

$$= 7500 \text{ kg/m}^3$$

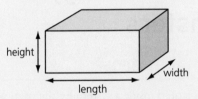

Figure 20.3 The volume of a cuboid = length × width × height.

The density of an irregular solid

Use the information in Figure 20.4 to calculate the density of the rock.

volume of the rock = 20 × 10⁻⁶ m³

$$\rho = \frac{0.09}{20 \times 10^{-6}}$$

$$= 4500 \text{ kg/m}^3$$

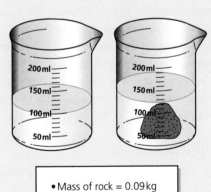

• Mass of rock = 0.09 kg

• 1 ml = $\dfrac{1}{1\,000\,000}$ m³

Figure 20.4

Now test yourself

1 Copy the table and fill in the gaps.

Material	Mass in kg	Volume in m³	Density in kg/m³
A	1800	4.5	
B	0.064		0.08
C		0.01	9000
D	600	0.03	

2 A cube of wood has a side length of 5.7 cm; the mass of the cube is 144.6 g. Calculate the density of the wood in kg/m³. Express your answer to an appropriate number of significant figures.
3 Explain how you would measure the density of an irregularly shaped solid. Include a description of the apparatus you would use and the measurements you would take. Explain what errors might occur in your experiment, and how you would attempt to reduce them.
4 (a) Draw diagrams to show the arrangements of atoms in the three states of matter.
   (b) Use your diagrams to explain why solids are much denser than gases.

Answers online

Internal energy and energy transfers

Internal energy

Energy is stored inside a system by the particles (atoms or molecules) that make up the system. This is called internal energy. The internal energy is the sum of the kinetic and potential energies of the particles that make up the system.

Heating

Heating increases the energy stored within a system by increasing the internal energy of the particles in the system.
- Heating can increase the temperature of the system – the atoms of the system move faster and the kinetic energy of the atoms rises.
- Heating can cause a change of state – for example, when a liquid evaporates to become a gas. The atoms increase their separation when the substance changes from a liquid to a gas, and this causes the atoms to increase their potential energy. So, the internal energy of the substance increases.

Changes of state

There is an increase in internal energy for:
- melting – a solid turns to a liquid
- boiling or evaporation – a liquid turns to a gas.

There is a decrease in internal energy for:
- freezing – a liquid turns to a solid
- condensation – a gas turns to a liquid.

Changes of state are physical changes and the process can be reversed. The change does not produce a new substance. There is no change in mass – we say that mass is conserved.

Now test yourself

5 (a) Explain what is meant by the term *internal energy*.
   (b) Explain two ways in which heating can increase the internal energy of a substance.
6 (a) What is meant by a *change of state* of a substance?
   (b) Give two examples of a change of state of a substance.

Answers online

Temperature changes in a system and specific heat capacity

If the temperature of a system increases, the increase in temperature depends on the mass of the substance heated, the type of material and the energy supplied to the system.

The following equation applies:

change in thermal energy = mass × specific heat capacity × temperature change

$$\Delta E = mc\Delta\theta$$

The specific heat capacity of a substance is the amount of energy required to raise the temperature of one kilogram of the substance by one degree Celsius.

> change in thermal energy, ΔE, in joules, J
>
> mass, m, in kilograms, kg
>
> specific heat capacity, c, in joules per kilogram per degree Celsius, J/kg°C
>
> temperature change, $\Delta\theta$, in degrees Celsius, °C

Example

12 000 J of energy are supplied to 4.0 kg of a substance and the temperature of the substances increases by 20° C. Calculate the specific heat capacity of the substance.

Answer

$$\Delta E = mc\Delta\theta$$

$$12\,000 = 4 \times c \times 20$$

$$c = \frac{12\,000}{4 \times 20}$$

$$= 150\,\text{J/kg°C}$$

> **Typical mistake**
>
> Remember the unit of specific heat capacity is J/kg°C.

Now test yourself

7 (a) State the unit of specific heat capacity.
   (b) Explain the meaning of *specific heat capacity*.
8 Use the information in the table to answer the questions.

Substance	Specific heat capacity J/kg °C
Water	4200
Glass	630
Air	1000

   (a) A kettle contains 0.5 kg of water. Calculate the energy required to warm the water from 20 °C to 70 °C.
   (b) A glass dish has a mass of 0.2 kg; 12 600 J of thermal energy is transferred to the dish. Calculate the temperature rise of the dish.
   (c) The mass of air in a room is 75 kg. Calculate the energy required to warm the air from a temperature of −5 °C to 20 °C.

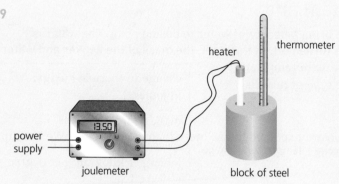

Figure 20.5

(a) The diagram shows the apparatus used to measure the specific heat capacity of steel. The mass of the block is 0.8 kg; the temperature of the block rises from 21 °C to 55 °C. Use the information in the diagram to calculate the specific heat capacity of steel.

(b) Explain why such an experiment is likely to give an answer for the specific heat capacity that is higher than the true value.

Answers online

Latent heat

When a pan of water is put onto a cooker, it heats up and will reach its boiling point at 100 °C. If the pan is left on the cooker, it will continue to boil, but there is no further increase in the water's temperature. Now the energy from the cooker is being used to change the state of the water. The energy increases the internal energy of the steam – as the molecules of water are separated, their potential energy increases. The internal energy of one kilogram of steam at 100 °C is greater than the internal energy of water at 100 °C.

The energy required to change the state of 1 kg of a substance, without a change of temperature, is called the specific latent heat.

energy for a change of state = mass × specific latent heat

$$E = mL$$

There are three states of matter, so each substance has two specific latent heats:

- The **specific latent heat of fusion** is the energy required to turn 1 kg of a solid into 1 kg of a liquid at the same temperature.
- The **specific latent heat of vaporisation** is the energy required to turn 1 kg of a liquid into 1 kg of a vapour at the same temperature.

> **Exam tip**
>
> When a substance changes from a solid to a liquid, the energy supplied increases the internal energy of the substance, but does not increase the temperature.

> energy, E, in joules, J
>
> mass, m, in kilograms, kg
>
> specific latent heat, L, in joules/kilogram, J/kg

Melting and freezing

When a substance melts, energy is supplied to increase the internal energy of the atoms (or molecules). When a substance freezes, the internal energy of the atoms reduces, and energy is released to the surroundings.

Evaporation, condensation and sublimation

When a substance evaporates, energy is supplied to increase the internal energy of the atoms (or molecules). When a substance condenses, the internal energy of the atoms reduces, and energy is released to the surroundings.

> **Sublimation** is a phase change of a substance directly from a solid to a vapour without passing through a liquid phase.

An immersion heater of power 500 W is used to bring a beaker of water to boiling point. The water is allowed to boil for 5 minutes, then the heater is turned off. At the start, the mass of the beaker and water was 632 g, at the end of the mass was 572 g.

Calculate the specific latent heat of vaporisation of water.

Answer

This is an example of a difficult problem and you are expected to use more than one equation. We have to work out E and m first before we use this equation:

$$E = mL$$

To find energy, E:

$$E = \text{power} \times \text{time}$$
$$= 500 \times 5 \times 60$$
$$= 150\,000\,\text{J}$$

> **Exam tip**
>
> Remember to turn 5 minutes into 300 seconds. Remember to turn 60 g into 0.06 kg.

To find the mass vaporised, m:

$$m = 632 - 572 = 60\,\text{g} = 0.06\,\text{kg}$$

Now using

$$E = mL$$
$$150\,000 = 0.06 \times L$$
$$L = \frac{150\,000}{0.06}$$
$$= 2\,500\,000\,\text{J/kg or } 2.5\,\text{MJ/kg}$$

> **Exam tip**
>
> The unit of specific latent heat is joules/kilogram, J/kg.

Cooling graphs

When a beaker of water cools down, the temperature changes as shown in Figure 20.6. The graph shows a steady drop in temperature, with the rate of cooling slowing down at lower temperatures.

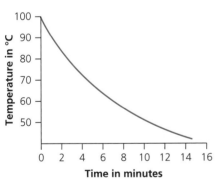

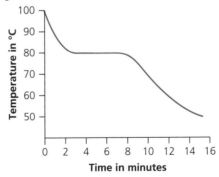

Figure 20.6 The cooling curve for water.

Figure 20.7 The cooling curve for ethanamide.

When a beaker of ethanamide cools, the cooling curve is very different – see Figure 20.7.

This shows us that ethanamide solidifies (or freezes) at a temperature of 80 °C. The beaker continues to transfer energy to the surroundings, although the temperature stays the same. As the molecules go from the liquid to the solid state, their internal energy decreases, and this allows the energy to be transferred from the ethanamide to the surroundings.

10 (a) Explain what is meant by the *specific latent heat of fusion* for a solid.
   (b) Give the unit of specific latent heat.
11 (a) Explain why sweating helps us cool down.
   (b) When a dog gets hot it pants. How does panting help the dog to cool?
   (c) A gardener places a large bucket of water in his greenhouse when the weather forecast predicts a severe frost. Explain how the bucket of water can protect the plants from the frost.
12 (a) Figure 20.8 shows a cooling curve for a hot substance.

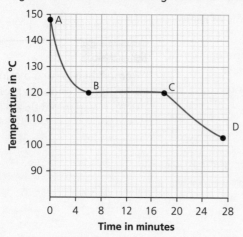

Figure 20.8

   (i) Explain why the substance cools more quickly over the region AB than it does over the region CD.
   (ii) At what temperature does the substance solidify?
   (b) A student writes the following statement.

   'The substance transfers energy to the surroundings at a faster rate at time B than it does at time C.'

   Explain whether this is true.
   (c) Explain whether it is possible to heat a substance without causing the temperature of the substance to increase.
13 A student sets up some apparatus to measure the specific latent heat of fusion of ice.
   The heater has a power of 75 W, and melts 24 g of ice in the 2 minutes.
   (a) Calculate the energy transferred to the ice by the heater in 2 minutes.
   (b) Use the data above to calculate the specific latent heat of fusion of ice. Give your answer in J/kg.
   (c) (i) Explain why it is important that the ice is allowed to reach a temperature of 0 °C, before the heater is switched on.
       (ii) State what type of error is introduced if the ice is warmed up from a temperature of −5 °C.
   (d) Explain what safety precautions you would take in planning this experiment.

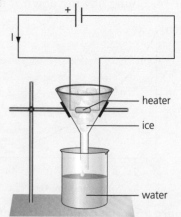

Figure 20.9

Answers online

Particle model and pressure

The particle model of gases

REVISED

The particle model helps us to understand the behaviour of gases. The main points of the model are:
- The particles in a gas (atoms or molecules) are in a constant state of random motion.
- The particles in a gas collide with each other and the walls of their container without losing any kinetic energy.
- The temperature of the gas is related to the average kinetic energy of the particles.
- As the average kinetic energy of the molecules increases, the temperature of the gas increases.

Gas pressure

- When the particles of a gas hit the wall of their container, the particles exert a force. In Figure 20.10 each of the molecules exerts a force at right angles to the wall as it bounces off the wall.

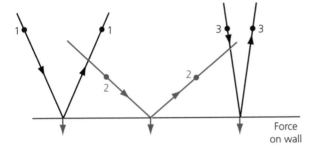

Figure 20.10

- The pressure inside a container of gas, with a fixed volume, increases when the temperature increases. At a higher temperature, the molecules move with greater speed. So they hit the walls of the container harder and more often. The force exerted on the walls increases and the pressure increases.
- The pressure inside a container of gas, at a constant temperature, decreases when the volume is increased. The molecules continue to move with the same average kinetic energy (and therefore speed), but the particles hit the walls less frequently. So the force exerted on the wall and, therefore, the pressure decreases.

Expanding and compressing gases

REVISED

Figure 20.11 shows apparatus that can be used to change the pressure and volume of a fixed mass of air at a constant temperature.

Figure 20.12 shows the relationship between the pressure and the volume of air in the column of gas in Figure 20.11.

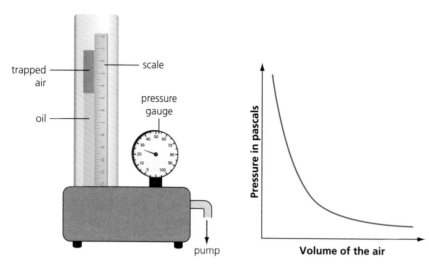

trapped air

scale

oil

pressure gauge

pump

Figure 20.11

Pressure in pascals

Volume of the air

Figure 20.12

By increasing the pressure on the oil, using a pump, the column of trapped air is compressed.

For a fixed mass of gas held at constant temperature:

pressure × volume = constant

$$PV = \text{constant}$$

pressure, P, in pascals, Pa

volume, V, in metres cubed, m^3

There is an inverse proportion between the volume and the pressure of a gas.

$$P \propto \frac{1}{V}$$

For example:

- halving the volume of the gas doubles the pressure
- increasing the volume of the gas by a factor of three, reduces the pressure to a third of the original value.

Example

A gas in a container of volume $2.0\,m^3$ has a pressure of $500\,kPa$. The volume is reduced to $0.40\,m^3$. Calculate the new pressure.

Answer

$$P_1V_1 = P_2V_2$$

$$500 \times 2 = P_2 \times 0.4$$

$$P_2 = \frac{500 \times 2}{0.4}$$

$$= 2500\,kPa$$

Now test yourself

14 (a) Describe the motion of the particles in a gas.
   (b) How does the motion of the particles change when the gas gets hotter?
15 Use the particle model of gas to explain:
   (a) how a gas exerts a pressure on the walls of its container
   (b) why the pressure of a gas, which is kept at constant volume, increases when its temperature is increased.

16 A balloon has a volume of 0.02 m³ when the pressure inside it is 100 kPa. A weight is attached to the balloon and it falls to the bottom of a lake where the pressure is 500 kPa. Calculate the volume of the balloon now. Assume the temperature of the balloon doesn't change.

17 A cylinder of gas has a volume of 0.40 m³ at a pressure of 700 kPa. Calculate the volume of the gas when the valve on the cylinder is opened and the gas is allowed to escape and reach atmospheric pressure of 100 kPa. Assume the temperature of the gas doesn't change.

Answers online

Summary

- Density is measured in kg/m³.

  density = mass/volume

- There are three states of matter: solid, liquid and gas.
  - The particles in a solid are closely packed and vibrate about fixed positions.
  - The particles in a liquid are in close contact, but are free to move past each other, so a liquid flows.
  - The particles in a gas are far apart, and they move randomly in all directions.
- The internal energy of a system of particles is the sum of the kinetic energy and potential energy of the particles. Heating increases the internal energy of the particles.
- Heating can raise the temperature of a substance. The particles move faster as the temperature rises.
- Heating can also cause a change of state, without changing the temperature of a substance.
- When heating causes a change of temperature the following equation applies:

  change in thermal energy = mass × specific heat capacity × temperature change

  $$\Delta E = mc\Delta\theta$$

  c is measured in J/kg °C

- When heating causes a change of state, the following equation applies.

  energy for a change of state = mass × specific latent heat

  $$E = mL$$

  L is measured in J/kg
- When a liquid cools, its temperature remains constant as it solidifies (see Figure 3.7).
- The particle model of gases states that gas molecules (or atoms) are in a constant state of random motion. Particles hit the walls of their container and exert a pressure. The pressure increases at higher temperatures; because the particles have greater kinetic energy, they hit the walls faster and more often.
- For a gas at constant temperature:

  pressure × volume = constant

  $$PV = \text{constant}$$

- The temperature of a gas can be increased by doing mechanical work to compress it.

Exam practice

1 Which of the following is the correct unit for specific heat capacity?
  A J/kg   B J kg °C   C J/kg °C   D J   [1]

2 The diagrams show the arrangements of particles in a solid and a gas.

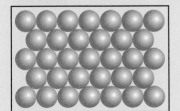 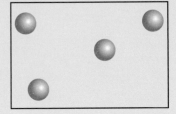

Figure 20.13

(a) (i) Describe the motion of the particles in the solid at room temperature. [1]
   (ii) Explain what happens to the particles when the temperature rises. [1]
(b) (i) Describe the motion of the particles in the gas. [1]
   (ii) Explain why the particles in a gas exert a pressure on the walls of their container. [2]
   (iii) Explain why the pressure exerted by the gas increases when the temperature of the gas rises. [2]
(c) A mixture of ice and water is put into a pan at a temperature of 0 °C. The pan is heated until all the ice melts.
   (i) Draw a diagram to show the arrangement of particles in a liquid. [1]
   (ii) Explain why the temperature of the ice and water remains at 0 °C until all the ice is melted, even though the mixture is being heated. [3]

3 A student uses a ruler to measure the side lengths of a cuboid of aluminium as shown in Figure 20.14. Aluminium has a density of 2700 kg/m³.

(a) Calculate the volume of the aluminium in m³. [2]
(b) Use the density of aluminium to calculate its mass. [2]
(c) When the student puts the aluminium onto an electronic balance, he records a mass of 205.8 g. Explain what might have caused the difference between the actual and calculated value of the mass. [2]

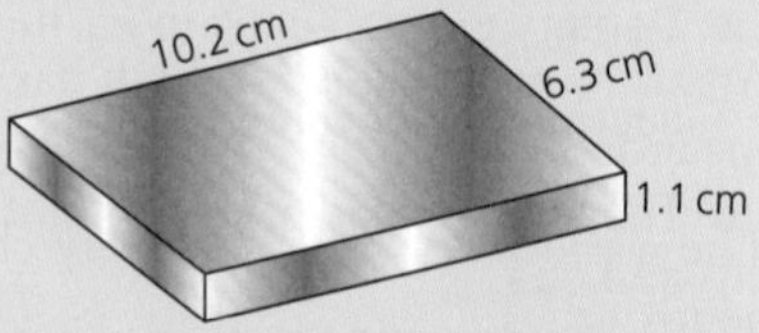

Figure 20.14

10.2 cm   6.3 cm   1.1 cm

4 A heater is used to heat a metal block of mass 0.5 kg. After the heater is turned on the temperature rises, as shown in Figure 20.15.

(a) Use the graph to determine the temperature rise in the first 60 s of heating. [1]
(b) During the first 60 s of heating 5025 J of energy is supplied to the block. Calculate the specific heat capacity of the block. [3]
(c) Use the information in part (b) to calculate the power of the heater. [2]

5 A small heater is placed into some crushed ice, and turned on for 4 minutes. The heater has a power of 36 W.
(a) Show that the heater transfers 8600 J to the ice in 4 minutes. [2]
(b) The specific latent heat of fusion of ice is 330 kJ/kg. Calculate the mass of ice melted after 4 minutes. [3]

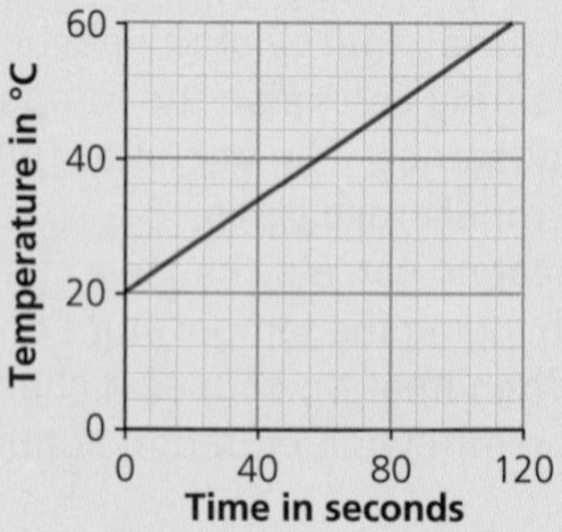

6 Some air is trapped in an airtight cylinder. A piston slowly compressed the air, so that the length of the air column is reduced from 40 cm to 15 cm. The air temperature remains constant. The initial air pressure, at a length of 40 cm, is 120 kPa.
(a) Explain in terms of the particle model of gases why the pressure of the air increases as the air column is compressed. [2]
(b) (i) Calculate the air pressure when the column has a length of 15 cm. [3]
   (ii) The cross-sectional area of the cylinder is 0.06 m² and the mass of gas in the cylinder is 0.02 kg. Calculate the density of the gas when the length of the air column is 15 cm. Express your answer in kg/m³. [3]

Figure 20.15

7 The air in a room is to be heated from 5 °C to 18 °C by a convector heating. Use the information in the list below to calculate the energy required to heat the air. [6]
   – Volume of air in the room: 70 m³
   – Density of air: 1.2 kg/m³
   – Specific heat capacity of air: 1000 J/kg °C

Answers and quick quiz 20 online

ONLINE

21 Atomic structure

Atoms and isotopes

The structure of an atom

Atoms are very small, having a radius of about 1×10^{-10} m.

An atom has a nucleus which has a radius less than 1/10 000 of the atom.

The nucleus contains positively charged protons and neutral neutrons.

Negatively charged electrons are arranged at different distances from the nucleus, which correspond to different energy levels. Electrons can change energy levels by the absorption or emission of electromagnetic radiation.

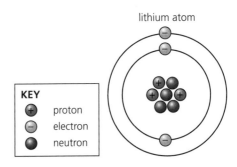

lithium atom

KEY
⊕ proton
⊖ electron
● neutron

Figure 21.1 The arrangement of protons, neutrons and electrons in a lithium atom. Note this is not drawn to scale.

Atoms and ions

In an atom the number of electrons is equal to the number of protons. Atoms have no overall charge, because the size of the negative charge on an electron is the same size as the positive charge on a proton.

If an atom gains an electron, it becomes a negative ion. If the atom loses an electron, it becomes a positive ion.

Mass number, atomic number and isotopes

An atom is determined by the number of protons in its nucleus. The number of protons in an element is called its **atomic number**.

The total number of protons and neutrons in an atom is called its **mass number**.

An atom can be represented as shown in this example:

$$\text{mass number 27} \atop \text{atomic number 13}} \text{Al}$$

This symbol tells us that aluminium has 13 protons in its nucleus and 14 neutrons, making a total of 27 protons and neutrons.

Isotopes

Not all the atoms of an element have the same mass, for example one atom of aluminium might have a mass of 27 and another a mass of 26. Both atoms have 13 protons, but one has 13 neutrons and the other 14 neutrons. These are two **isotopes** of aluminium:

- $^{27}_{13}\text{Al}$ is the most common isotope of aluminium.
- $^{26}_{13}\text{Al}$ is another isotope of aluminium.

> **Atomic number** is the number of protons.
>
> **Mass number** is the number of protons and neutrons.
>
> **Isotopes** are different forms of a particular element. Isotopes have the same number of protons but different numbers of neutrons.

Now test yourself

1 (a) Which of the following is the approximate radius of an atom?

10^{-4} m $\quad$ 10^{-7} m $\quad$ 10^{-10} m

(b) Use one of the answers from the list below to complete the following sentence.

The radius of an atom is approximately .... times the radius of a nucleus.

100 $\quad$ 10 000 $\quad$ 10 000 000

2 An oxygen atom has 8 protons, 8 neutrons and 8 electrons.
(a) State the atomic number of oxygen.
(b) State the mass number of oxygen.
(c) Explain why an oxygen atom is neutral.

3 Calculate the number of protons and neutrons in each of the following nuclei:

(a) $^{11}_{5}\text{B}$

(b) $^{32}_{16}\text{S}$

(c) $^{156}_{64}\text{Gd}$

(d) $^{237}_{93}\text{Np}$

4 Explain the meaning of the following terms:
(a) atomic number
(b) mass number
(c) isotope
(d) ion

5 The atomic radius of uranium is about 1.8×10^{-10} m and the nuclear radius of uranium is about 7.5×10^{-15} m. Calculate the ratio of atomic to nuclear radii in uranium.

Answers online

The development of the model of the atom

New experimental evidence may lead to a scientific model being changed.

Until the discovery of the electron in 1897, atoms were thought to be indivisible solid spheres, and the smallest part of matter.

The 'plum pudding' model

In 1904, J J Thompson proposed a new model for the atom. The idea of the model was that an atom was made up of a positive ball of matter, with electrons dotted inside – the electrons are the plums in the pudding.

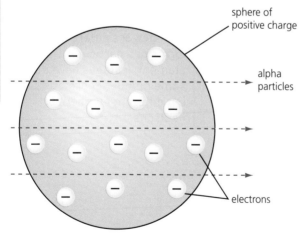

Figure 21.2 The 'plum pudding' atomic model.

The nuclear model of the atom

In 1909 the Geiger and Marsden experiment led to the idea of the nuclear atom. They directed a beam of alpha particles (He^{2+} nuclei) at a thin gold foil.

They expected the alpha particles to travel straight through as shown in Figure 21.2.

In fact, most alpha particles did travel straight through the foil. But a very small fraction of alpha particles bounced back, as shown in Figure 21.3.

The conclusion we now draw from the Geiger and Marsden experiment is that the alpha particles are repelled by a very small, positively charged nucleus, which contains most of the mass of the atom.

- The nucleus must be small because only a small fraction of alpha particles bounce back.
- The nucleus is positive because its strong electric field repels the positively charged alpha particles.
- The nucleus must be massive, because a small nucleus would be knocked forwards by the alpha particle.

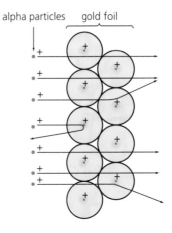

alpha particles   gold foil

Figure 21.3

The Bohr model of the atom

In 1913 Neils Bohr suggested a model of the atom in which electrons move round the nucleus in circular orbits. In this model electrons can change their orbit.

Scientists had discovered that matter absorbs and emits specific energies or electromagnetic radiation.

The Bohr model explains this:

- When an electron falls from a high level to a low level, it emits electromagnetic radiation – for example, falling from level 3 to 2 in Figure 21.4.
- An electron jumps up a level by absorbing electromagnetic radiation.

Later experiments showed that the nucleus could be divided further into protons and neutrons. James Chadwick discovered neutrons in 1932.

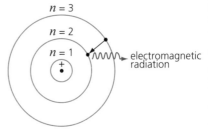

Figure 21.4

Now test yourself

TESTED ☐

6 Explain and describe the experimental evidence that led to the nuclear model of the atom.
7 Describe the Bohr model of the atom.

Answers online

Atoms and nuclear radiation

Some atomic nuclei are unstable. The nucleus gives out radiation and it changes to become more stable. This is a random process which is called radioactive decay.

The activity of a radioactive source is the rate at which it decays.

Activity is measured in Becquerel, Bq.

$1\,Bq = 1$ nuclear decay per second.

A small radioactive source might have a decay rate of $10^6\,Bq$.

Ionising radioactive particles may be detected using a Geiger–Müller (GM) tube.

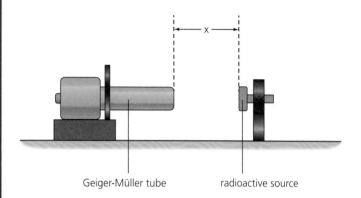

Geiger-Müller tube   radioactive source

Figure 21.5

The count rate detected by a GM tube is always less than the activity of a radioactive source, because the source emits particles in all directions.

Nuclear radiation

REVISED

Nuclear radiation that may be emitted from nuclei include:
- an alpha particle (α) – this consists of two protons and two neutrons, which is the same as a helium nucleus
- a beta particle (β) – this is a high-speed electron that escapes from a nucleus when a neutron turns into a proton
- a gamma (γ) ray – this is electromagnetic radiation emitted from the nucleus
- a neutron (n).

Properties of radiation

REVISED

- Alpha particles travel about 5 cm through air and can be stopped by a piece of paper. Alpha particles are strongly ionising.
- Beta particles can travel several metres through air and can be stopped by a sheet of aluminium that is a few millimetres thick. Beta particles are not as strongly ionising as alpha particles.
- Gamma rays can only be effectively stopped by very thick lead. Gamma rays are only weakly ionising and travel great distances in air.

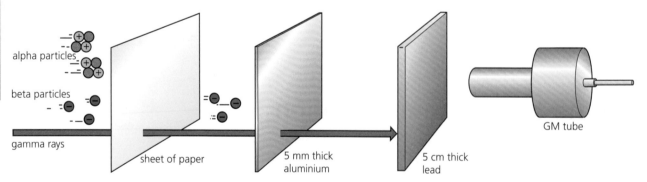

alpha particles

beta particles

gamma rays

sheet of paper   5 mm thick aluminium   5 cm thick lead   GM tube

Figure 21.6

Radiation	Nature	Range in air	Ionising power	Penetrating power
Alpha α	helium nucleus	a few centimetres	very strong	stopped by paper
Beta β	electron	a few metres	medium	stopped by aluminium
Gamma γ	electromagnetic waves	great distances	weak	stopped by thick lead

Answers and quick quizzes at **www.hoddereducation.co.uk/myrevisionnotesdownloads**

Radiation damage

Radiation that gets into our bodies can cause damage to our cells. Alpha particles cause the most damage – this could happen if we inhaled a radioactive gas.

Gamma rays are less ionising than alpha particles, but they can get into our body because they are very penetrating.

Nuclear equations

REVISED

An alpha particle may be represented by the symbol:

$$_2^4 He$$

So when an alpha particle is emitted from a nucleus, it causes the mass number to decrease by 4 and the atomic number by 2.

For example:

$$_{89}^{225} Ac \rightarrow _{87}^{221} Fr + _2^4 He$$

A beta particle may be represented by the symbol:

$$_{-1}^0 e$$

So when a beta particle is emitted from a nucleus, the mass number remains the same but the atomic number increases by 1.

For example:

$$_1^3 H \rightarrow _2^3 He + _{-1}^0 e$$

The emission of a gamma ray from a nucleus does not cause the mass or atomic number to change. The gamma ray has no mass or charge, but it does carry away some energy from the nuclear store.

> **Typical mistake**
>
> Remember that when a beta particle is emitted, the atomic number **increases** by 1; it does **not** decrease.

Now test yourself

TESTED

8 Which of the following are properties of beta radiation?
  A It is the most strongly ionising radiation.
  B It is stopped by aluminium.
  C It is a fast-moving electron.
9 Explain why a teacher uses long tongs when she handles radioactive sources.
10 (a) Explain what is meant by the *activity* of a radioactive source.
   (b) Which of the following is the correct unit for the activity of a radioactive source?
      rutherford   geiger   becquerel
11 Explain the nature of each of the following:
   (a) an alpha particle
   (b) a beta particle
   (c) a gamma ray.
12 When a gamma ray is emitted from a nucleus, what changes occur to the mass and atomic numbers of the nucleus?
13 Fill in the gaps in the following radioactive equations:

(a) $_{94}^{241} Pu \rightarrow _{92}^{?} U + _?^4 He$

(b) $_{92}^{237} U \rightarrow _?^{237} Np + _{-1}^0 e$

(c) $_{26}^{?} Fe \rightarrow _{27}^{59} Co + _?^? e$

(d) $_{84}^{213} Po \rightarrow _?^? Pb + _2^4 He$

(e) $_{14}^{32} Si \rightarrow _{15}^{32} P + ?$

(f) $_{90}^{229} Th \rightarrow _{88}^{225} Ra + ?$

Answers online

Half-lives and the random nature of radioactive decay

Radioactive decay is random.

We use the analogy of rolling lots of dice to help explain this. If you roll one die, it is not possible to predict if you will throw a six. If you throw 600 dice, on average you would expect to throw about 100 sixes. Radioactive decay is like that: you cannot predict when one nucleus will decay, but when there are lots of nuclei they decay in a predictable way.

Half-life

REVISED

The half-life of a radioactive isotope is the time it takes for the number of nuclei in a sample to halve. The half-life is also the time it takes for the count rate (or activity) detected by a GM tube (or other detector) to halve.

The half-life is a measure of the stability of a nucleus. A shorter half-life means the nucleus is less stable.

Measuring half-lives

You can determine the half-life by measuring the activity of a radioactive isotope. The graph shows how the measured count rate for an isotope changes with time. You can see that the half-life for this isotope is 50 seconds. Every 50 seconds the count rate halves.

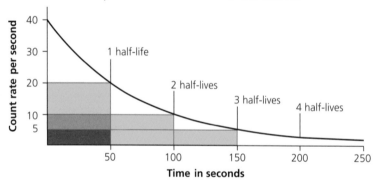

Figure 21.7

Example

A radioactive isotope has an activity of 1.6×10^6 Bq. The half-life of the isotope is 8 hours. Calculate the activity of the isotope after a day.

Answer

One day is 24 hours which is three half-lives.

So the activity will be

$$\frac{1}{2} \times \frac{1}{2} \times \frac{1}{2} \times 1.6 \times 10^6 \text{ Bq}$$

$$= 2 \times 10^5 \text{ Bq}$$

Exam tip

The activity of a radioactive isotope is:

- $\frac{1}{2}$ after 1 half-life
- $\frac{1}{4}$ after 2 half-lives
- $\frac{1}{8}$ after 3 half-lives
- $\frac{1}{16}$ after 4 half-lives.

Answers and quick quizzes at **www.hoddereducation.co.uk/myrevisionnotesdownloads**

Now test yourself

14 Use a word from the list to complete the sentence.

count rate   activity   reaction

The .... is the number of particles emitted in one second by a radioactive source.

15 Explain what the word *random* means.

16 The graph Figure 21.8 shows the decay of three different radioactive isotopes.
   Which isotope:
   (a) has the longest half-life
   (b) has the shortest half-life

17 A scientist measures the count rate for a radioactive isotope. His measurements are shown in the table.

Count rate Bq	Time in hours
12 000	0
9240	2
7110	4
5480	6
4220	8
3250	10
2500	12

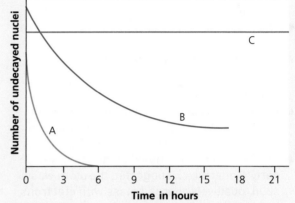

Figure 21.8

Plot a graph to determine the half-life of the isotope.

18 An isotope has a half-life of 16 days.
   Today its activity is measured to be 4.0×10^5 Bq.
   Calculate its activity in 64 days' time.

Answers online

Radioactive contamination

There is an important difference between **irradiation** and radioactive **contamination**.

● A patient may be exposed to radiation in a course of radiotherapy. Then the body is irradiated by a specific dose of radiation. Once the radioactive source is removed, the patient's body is not radioactive.

● Radioactive contamination occurs when unwanted radioactive material is absorbed by another material. For example, if there is a leak of radioactive waste from a power station, radioactive material may flow into a river. Then animals that drink from the river absorb radioactive materials into their bodies. In this way an animal is contaminated and continuously exposed to radiation.

> **Exam tip**
>
> Make sure you know the difference between irradiation and contamination.

Screening from radiation

When a radioactive source is used to irradiate something, the operator takes precautions to avoid exposure to radiation. For example, radiographers in hospitals wear lead aprons, and keep well away from a source when it is being used.

Radioactive contamination provides a great risk to us, as we cannot shield ourselves – if we have absorbed a contaminated material, the radioactive source is inside our body.

Summary

- An atom has a radius of about 10^{-10} m.
- The radius of a nucleus is less than 1/10000 of the radius of an atom.
- A nucleus contains positively charged protons and neutral neutrons. A neutral atom contains as many negatively charged electrons as protons.
- The atomic number of an atom is the number of protons.
- The mass number of an atom is the sum of the numbers of protons and neutrons.
- These numbers may be represented:

$$^{\text{mass number } 16}_{\text{atomic number } 8}\text{O}$$

- An element is determined by the number of protons in the nucleus. Different isotopes of an element have different numbers of neutrons.
- The 'plum pudding' model of the atom was an early model which suggested the atom was a solid, positively charged mass with electrons inside.
  The scattering of alpha particles led to the nuclear model of the atom, with a small, massive, positively charged nucleus surrounded by electrons.
- The Bohr model of the atom stated that electrons orbit the nucleus and can change their energy level by absorbing or emitting electromagnetic radiation.
- An alpha particle is a helium nucleus. It is strongly ionising, travels about 5 cm in air and is stopped by paper. It is represented by:

$$^4_2\text{He}$$

- A beta particle is a fast electron. It is less ionising than an alpha particle and is stopped by aluminium a few millimetres thick. It is represented by:

$$^{\ 0}_{-1}\text{e}$$

- A gamma ray is an electromagnetic ray. It is weakly ionising and only stopped by very thick lead.
- Alpha emission reduces the mass number of a nucleus by 4 and the atomic number by 2. For example:

$$^{219}_{86}\text{Rn} \rightarrow {}^{215}_{84}\text{Po} + {}^4_2\text{He}$$

- Beta emission increases the atomic number of a nucleus by 1 and leaves the mass number unchanged. For example:

$$^{14}_{6}\text{C} \rightarrow {}^{14}_{7}\text{N} + {}^{\ 0}_{-1}\text{e}$$

- Radioactive decay is random.
- Radioactive decay has a half-life. For every half-life that passes, the activity of a radioactive source and the number of radioactive nuclei halves.
- The activity of a radioactive source is the number of emissions per second. This is measured in becquerel, Bq.
- Radioactive contamination occurs when unwanted radioactive materials are present in other materials. Irradiation occurs when a material is exposed to an external source of radiation.

Exam practice

1 A radioactive source emits alpha (α), beta (β) and gamma (γ) radiation.
   (a) Which of the radiations is most ionising? [1]
   (b) Which two radiations will pass through a piece of paper? [1]
   (c) Which radiation has the greatest range in air? [1]
2 (a) Fresh raspberries are sometimes irradiated before being transported to the UK. The irradiation kills bacteria on the raspberries.
      Which of these statements is true?
      A The raspberries become radioactive and cannot be eaten for a week.
      B The irradiation contaminates the raspberries.
      C Radioactive particles settle on the raspberries.
      D The raspberries do not become radioactive, and are safe to eat. [1]
   (b) Suggest a reason why a farmer would want to irradiate his raspberries. [1]

→

3 Figure 21.9 represents an atom of boron-11.
   (a) State the number of
      (i) protons
      (ii) neutrons
      (iii) electrons in the atom. [3]
   (b) State the atomic number of boron. Give a reason
      for your answer. [2]
   (c) Boron-12 is a radioactive isotope of boron.
      (i) Explain the word *isotope*. [1]
      (ii) How does boron-12 differ from boron-11? [1]
   (d) Boron-12 decays by the emission of a beta particle.
      (i) Which of the following describes a beta particle?
         A a helium nucleus
         B an electron from the nucleus
         C an electromagnetic wave [1]
      (ii) Complete the following equation that describes the decay of boron-12. [2]

   $$^{12}_{5}B \rightarrow \,^{?}_{?}C + \,^{?}_{?}\text{beta}$$

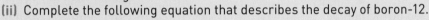

Figure 21.9

4 Gadolinium-148, $^{148}_{64}Gd$, is a radioactive isotope that decays by emitting alpha particles.
   (a) (i) State the atomic number of gadolinium [1]
      (ii) Calculate the number of neutrons in a nucleus of gadolinium-148. [1]
   (b) Complete the following equation that describes the decay of gadolinium-148. [2]

   $$^{148}_{64}Gd \rightarrow \,^{?}_{62}Sm + \,^{4}_{?}He$$

   (c) The graph shows how the activity of a sample of gadolinium-148 changes over a period of time.

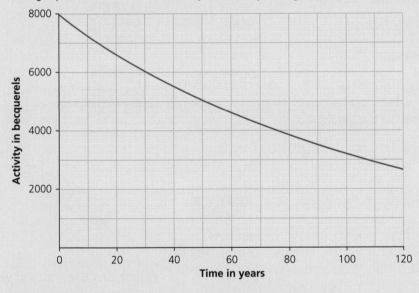

Figure 21.10

      (i) Explain what is meant by the word *activity*. [1]
      (ii) Calculate the half-life of gadolinium-148. [2]
      (iii) Use your answer to (ii) to predict how long it will take for the activity to drop
         from 8000 Bq to 1000 Bq. [2]

Answers and quick quiz 21 online

ONLINE

22 Forces

Forces and their interactions

Scalars and vectors

REVISED

Scalar quantities have magnitude (size) only. Examples include:
- speed
- mass
- distance
- energy.

Vector quantities have both magnitude (size) and direction. Examples include:
- force
- acceleration
- velocity.

We represent vectors with an arrow; the direction of the arrow shows the direction of the vector and the length of the arrow the magnitude of the quantity.

> **Exam tip**
>
> Velocity and speed are often used to mean the same thing. But speed is a scalar (e.g. 15 m/s) and velocity is a vector, which has both magnitude and direction (e.g. 15 m/s to the right).

Now test yourself

TESTED

1 Which of the following quantities are vector quantities?
force   mass   distance   acceleration
speed   energy   velocity

2 What is wrong with the following statement?

'A force of 3 N acts on an object.'

3 Car A travels due north on a motorway at 30 m/s. Car B travels due south on the motorway at 15 m/s.

Draw vectors to represent these two velocities.

Answers online

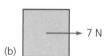

(a)

(b)

(c)

Figure 22.1 Examples of vectors.

Contact and non-contact forces

REVISED

- Contact forces act when one body touches another. Examples of contact forces include: friction, air resistance, tension in a rope, and the normal contact force when one object rests against another.
- When non-contact forces act, bodies are physically separated. Gravitational, electrostatic and magnetic forces are non-contact forces.

Gravity

REVISED

Weight is the force that acts on an object due to gravity.

The force of gravity around the Earth is due to the gravitational field around the Earth.

The weight of an object depends on the strength of the gravitational field. Different planets have different gravitational field strengths near their surfaces.

The weight of an object is calculated using the equation:

weight = mass × gravitational field strength

$$W = mg$$

weight, W, in newtons, N

mass, m, in kilograms, kg

gravitational field strength, g, in newtons per kilogram, N/kg

Resultant forces

When two or more forces act on an object, those forces may be replaced by a single force that has the same effect as all the original forces acting together.

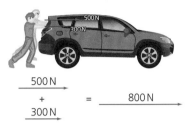

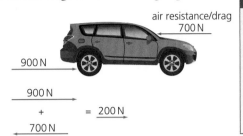

Figure 22.2 When two forces act in the same direction, they add up to make a larger force. This is called the resultant force.

Figure 22.3 When two forces act in different directions, a smaller resultant force is produced.

Freebody diagrams

A freebody diagram shows all the forces acting on a body.
In Figure 22.4 a man, pulling on a rope, has four forces acting on him:

- weight (W)   800 N down
- normal reaction force (R)   800 N up
- the tension from a rope (I)   150 N to the right
- friction (F)   50 N to the left

The resultant force on the man is: 100 N to the right.

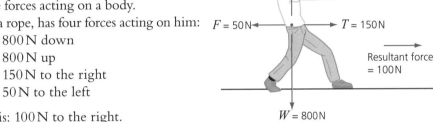

Figure 22.4

Resolving forces

A single force can be resolved into two components acting at right angles to each other.

A force of 10 N acts on a box as shown in Figure 22.5. This can be resolved into a vertical component of 6 N and a horizontal component of 8 N.

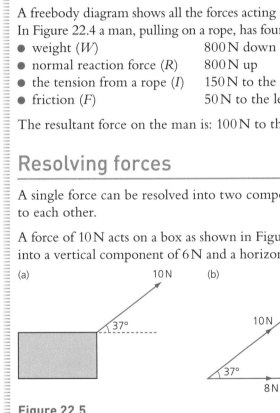

Figure 22.5

Figure 22.6 shows the direction and size of two forces exerted by two tugboats on a large ship. These are the forces of (i) 50 000 N along OA and (ii) 40 000 N along OB.

Determine by scale drawing the resultant of these two forces.

Answer

Complete the 'parallelogram' of forces by marking in AC (parallel to OA). The resultant force is the red line OC. Using the scale 1 cm = 10 000 N, the resultant force has a length of 7.5 cm on the diagram or 75 000 N.

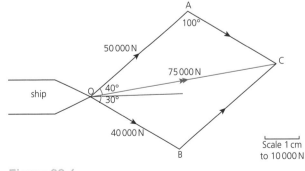

Figure 22.6

TESTED

4 Name three contact forces, and three non-contact forces.
5 State the units of:
  (a) weight
  (b) gravitational field strength.
6 An astronaut has a mass of 130 kg in his spacesuit. Calculate his weight on the Moon where the gravitational field strength is 1.6 N/kg.
7 Calculate the resultant forces on the two boxes in Figure 22.7.

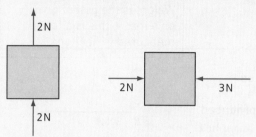

Figure 22.7

8 (a) In Figure 22.8(a), resolve the force of 35 N into a horizontal and a vertical component.

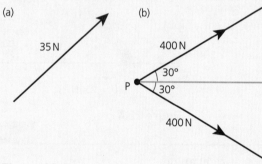

Figure 22.8

(b) In Figure 22.8(b), two forces of 400 N act on point P. Determine the resultant of these two forces by a scale drawing.

Answers online

Work done and energy transfer

When a force causes an object to move in the direction of the applied force, work is done. For example, when you drag a box along the floor, you do work against a frictional force.

work done = force × distance

$$W = Fs$$

One joule of work is done when a force of one newton causes a displacement of one metre.

1 joule = 1 newton-metre

Energy transfer

REVISED

When work is done on an object, energy is transferred from one store to another.

For example: when you lift a weight, energy is transferred from the chemical store in your arm to the gravitational potential energy store of the weight (and also to the thermal store in your arm).

work done, W, in joules, J

force, F, in newtons, N

distance, s, in metres, m

Typical mistake

When you hold a weight, without moving it, you get tired; you are transferring energy from your chemical store to a thermal store. But you are not doing any work – you only do work when you move the weight.

Answers and quick quizzes at **www.hoddereducation.co.uk/myrevisionnotes**

TESTED

Now test yourself

9 State the unit of work.
10 You hold a 20 N weight, without moving it, at arm's length for 5 minutes. Your arm gets tired.
   (a) Discuss what energy transfer takes place.
   (b) State how much work you do while holding the weight.
   (c) Calculate the work done when you lift the 20 N weight through a vertical height of 1.3 m.
11 (a) Calculate the work done when a force of 30 N is used to drag a bag 2.5 m along the floor.
   (b) Discuss the energy transfers in this process.

Answers online

Forces and elasticity

To bend, stretch or change the shape of an object you have to apply at least two forces. In Figure 22.9 a spring is stretched when a force is applied to each end.

If you only apply one force to a spring, you set it in motion but you do not stretch it.

The extension of a spring is directly proportional to the force applied, provided the limit of proportionality is not exceeded (point P on the graph).

force = spring constant × extension

$$F = ke$$

> force, F, in newtons, N
>
> spring constant, k, in newtons per metre, N/m
>
> extension, e, in metres, m

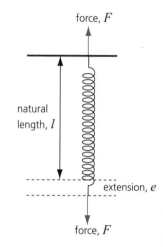

Figure 22.9

This relationship also applies to the compression of an elastic object, where e would be the compression of the object.

Required practical 18 investigates the relationship between force and extension of a spring. See Exam practice question 5 on page 236.

Elastic and inelastic deformation

REVISED

When a spring is stretched elastically, it returns to its original length and shape.

When a spring is stretched beyond the limit of proportionality, the spring is stretched inelastically. This means the spring does not return to its original length and shape.

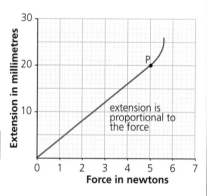

Figure 22.10

Energy transfers

REVISED

A force that stretches a spring does work and elastic potential energy is stored in the spring. Provided the spring is not deformed beyond the limit of proportionality, the work done in stretching the spring is equal to the elastic potential energy stored in the spring.

When a stretched spring is released, the stored elastic potential energy can be transferred to other energy stores, such as a kinetic store.

The elastic potential energy stored in a spring can be calculated as follows:

elastic potential energy = 0.5 × spring constant × extension²

$$E_e = \frac{1}{2}ke^2$$

> elastic potential energy, E_e, in joules, J
>
> spring constant, k, in newtons per metre, N/m
>
> extension, e, in metres, m

12 Explain the terms:
   (a) *elastic deformation*
   (b) *inelastic deformation.*
13 The graph in Figure 22.10 shows the extension of a spring for various applied forces.
   (a) Use the graph to determine how far the spring can be stretched before passing the limit of proportionality.
   (b) Calculate the spring constant in N/m.
   (c) Calculate the work done to stretch the spring by 20 mm.

Answers online

Describing motion along a line

Distance and displacement

REVISED

Distance is how far an object moves. Distance does not involve direction. Distance is a scalar quantity.

Displacement includes both the distance an object moves, measured in a straight line from the start point to the finish point and the direction of that straight line.

Displacement is a vector quantity.

Example

Figure 22.11 shows the path taken by a walker. She walks 8 km due east, (AB), then 6 km due south (BC).

The distance walked is 8 km + 6 km = 14 km.

The displacement is shown by the vector AC. This is 10 km along a direction 37° south of east.

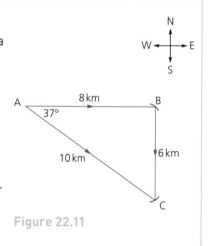

Figure 22.11

Speed and velocity

REVISED

For an object moving at a constant speed, the distance travelled in a specific time can be calculated using the equation:

distance travelled = speed × time

$$s = vt$$

As speed does not involve direction, it is a scalar quantity.

Velocity is the speed of an object in a given direction. Velocity is a vector quantity.

distance, s, in metres, m

speed, v, in metres per second, m/s

time, t, in seconds, s

Typical speeds

The speed at which a person can walk, run or cycle depends on their age, fitness, how far they have already travelled and the nature of the ground being travelled over.

Typical values may be taken as:
- walking   1.5 m/s
- running   3 m/s
- cycling   6 m/s

Ⓗ Motion in a circle

Figure 22.12 shows a satellite in a circular orbit around the Earth. The satellite moves at a constant speed, but its velocity changes all the time. The velocity changes because the direction of travel changes. The pull of the Earth's gravity changes the satellite's direction of travel.

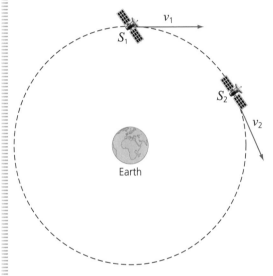

Figure 22.12

Distance–time graphs

When an object moves along a straight line, we can represent how far it has travelled by a distance–time graph.

Figure 22.13 shows a distance–time graph for a runner.

The speed of the runner is calculated using the gradient of the graph.

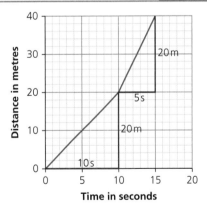

Figure 22.13

Example

Using the graph in Figure 22.13, calculate the speed of the runner between 10 and 15 seconds.

Answer

The speed from 10 s to 15 s is:

$$\text{speed} = \frac{\text{distance}}{\text{time}}$$

$$= \frac{20}{5}$$

$$= 4 \text{ m/s}$$

When an object is accelerating, the gradient of the distance–time graph changes.

We calculate the gradient by drawing a tangent to the curve.

In Figure 22.14, the gradient at A is:

$$\text{speed} = \frac{\text{distance}}{\text{time}}$$

$$= \frac{75}{25}$$

$$= 3\,\text{m/s}$$

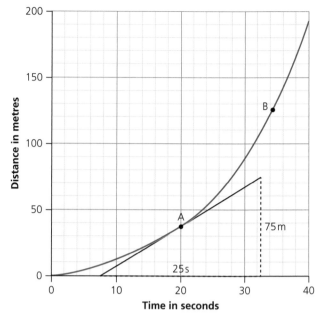

Figure 22.14

Now test yourself

14 An athlete runs (i) 100 m in 10.0 s and (ii) 400 m in 46.0 s.
  (a) Calculate his speed in each case.
  (b) Explain why one speed is faster than the other.
15 Figure 22.15 shows a distance–time graph for a car travelling along a straight road.

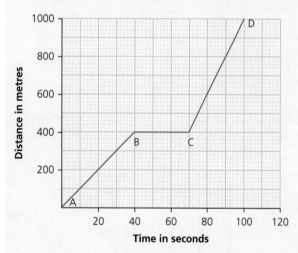

Figure 22.15

  (a) How far did the car travel over the region CD?
  (b) Over which part of the journey did the car travel at its greatest speed? Explain your answer.
  (c) For how long was the car stopped?
  (d) Calculate the car's speed over the region AB.
16 Calculate the speed of the moving object at point B in the distance–time graph in Figure 22.14.
17 A train travels at a constant speed of 55 m/s.
  (a) Calculate the distance travelled by the train in 1 minute.
  (b) Calculate the time it takes the train to travel 11 km.

Answers online

Acceleration

The average acceleration of an object is calculated using the equation:

$$acceleration = \frac{change\ in\ velocity}{time\ taken}$$

$$a = \frac{\Delta v}{t}$$

acceleration, a, in metres per second squared, m/s²

change in velocity, Δv, in metres per second, m/s

time, t, in seconds, s

When an object slows down it is decelerating.

Velocity–time graphs

Figure 22.16 shows a velocity–time graph for a cyclist.
- He accelerates up to 12 m/s over the first 8 seconds – section AB.
- He travels at a constant speed for the next 12 seconds – section BC.
- Then he decelerates from 12 m/s over the last 4 seconds – section CD.

Typical mistake

Often students give the unit of acceleration as m/s – it is **m/s²**. An acceleration is a change in velocity, m/s, in a given time, s. So the unit is m/s divided by s – **m/s²**.

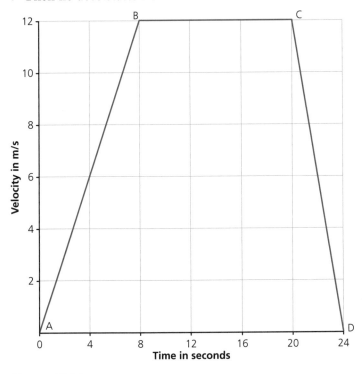

Figure 22.16

You can calculate the acceleration from a velocity–time graph.

Acceleration over time AB:

$$a = \frac{\Delta v}{t}$$
$$= \frac{12}{8}$$
$$= 1.5\,m/s^2$$

You can work out the distance travelled from a velocity–time graph.

Examples

Calculate how far the cyclist travelled:
1 when he travelled at a constant speed
2 while he accelerated.

Answers

1 $d = vt$

$\quad = 12 \times 12$

$\quad = 144\,m$

We can also work out the distance by counting the squares: under the line BC there are $6 \times 3 = 18$ squares.
But each square has a value $= 2\,m/s \times 4\,s = 8\,m$.
So the distance travelled $= 18 \times 8 = 144\,m$

2 $d =$ average speed $\times$ time

$\quad = 6 \times 8$

$\quad = 48\,m$

Note, here we have to use the average speed which is $6\,m/s$ – half-way between 0 and $12\,m/s$.
Or we could have worked out the distance using the 'area' under the graph:

'area' $= \dfrac{1}{2} \times 12\,m/s \times 8\,s = 48\,m$

This is equivalent to 6 squares.
In a Higher Tier exam question, you might be asked to work out the distance travelled for a graph with a curved line.

> **Exam tip**
>
> The gradient of a velocity–time graph equals the acceleration.

> **Exam tip**
>
> The area under a velocity–time graph equals the distance travelled.

Equation of motion for uniform acceleration

REVISED

The following equation applies to uniform acceleration:

(final velocity)^2 – (initial velocity)^2 = 2 × acceleration × distance

$$v^2 - u^2 = 2as$$

Example

A light aircraft accelerates from rest along a runway. The aircraft takes off at a speed of $35\,m/s$ having travelled $450\,m$ along the runway. Calculate the aircraft's acceleration.

Answer

$v^2 - u^2 = 2as$

$35^2 - 0 = 2 \times a \times 450$

$a = \dfrac{35^2}{900}$

$\quad = \dfrac{1225}{900}$

$\quad = 1.4\,m/s^2$

> final velocity, v, in metres per second, m/s
>
> initial velocity, u, in metres per second, m/s
>
> acceleration, a, in metres per second squared, m/s^2
>
> distance, s, in metres, m

Terminal velocity

When an object falls freely under gravity, near the Earth's surface, it has an acceleration of $9.8\,m/s^2$.

When an object falls through a **fluid**, resistive forces act on the object. Initially the acceleration due to gravity is $9.8\,m/s^2$. But as the object increases its speed, larger resistive forces act on it; so the resultant force on the object decreases and the acceleration decreases. Eventually the resultant force is zero and the object moves at a constant velocity. This is called the **terminal velocity**.

> A **fluid** is something that flows – either a gas or a liquid.
>
> An object reaches its **terminal velocity** when its weight is balanced by resistive forces.

Now test yourself

18 State the correct units for
 (a) velocity
 (b) acceleration.
19 (a) A runner accelerates from a speed of 3 m/s to 5 m/s in a time of 8 seconds. Calculate her acceleration.
 (b) A car slows down from a speed of 30 m/s to a speed of 18 m/s in a time of 3 seconds. Calculate the deceleration of the car.
20 This question refers to the velocity–time graph in Figure 22.19.
 (a) Calculate the deceleration of the cyclist in the last 4 seconds of his journey.
 (b) Calculate the distance travelled while the cyclist decelerates.
21 A train slows down from a speed of 50 m/s to 30 m/s over a distance of 1 km. Calculate the train's deceleration.

Answers online

> **Exam tip**
>
> You should be able to interpret a velocity–time graph in terms of the forces acting on an object falling and reaching its terminal velocity. Remember the gradient of the graph is the acceleration – so acceleration decreases as the velocity increases.

Forces, accelerations and Newton's laws of motion

Newton's first law

Newton's first law states that:

If the resultant force acting on an object is zero:
- and the object is stationary, the object remains stationary
- and when the object is moving, the object continues to move with the same speed in a straight line.

Newton's first law tells us that the velocity of an object only changes if a resultant force acts on an object. When a resultant force acts, a moving body can:
- speed up
- slow down
- or change direction.

When a car travels at a constant speed along a road, the resistive forces on the car balance the driving force on the car.

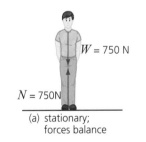

(a) stationary;
 forces balance

(b) moving at steady speed;
 forces balance

Figure 22.17

Inertia

The word *inertia* comes from the Latin word meaning *inactivity* or *inaction*.

In physics, we use the word *inertia* to describe an object's tendency to remain at rest or to continue moving at a constant speed. **Inertial mass** can be defined as the ratio of force over acceleration:

$$m = \frac{F}{a}$$

$$\text{Inertial mass} = \frac{\text{force}}{\text{acceleration}}$$

Newton's second law

The acceleration of an object is proportional to the resultant force acting on the object and inversely proportional to the mass of the object.

The statement above can be expressed mathematically in this form:

$$a \propto F$$

$$a \propto \frac{1}{m}$$

Newton's second law may also be written as an equation:

resultant force = mass × acceleration

$$F = ma$$

Typical mistake

When using Newton's second law make sure you have calculated the **resultant** force.

force, F, in newtons, N

mass, m, in kilograms, kg

acceleration, a, in metres per second squared, m/s²

Example

Figure 22.18

The mass of the car is 1000 kg; calculate its acceleration.

Answer

$$F = ma$$

$$800 - 500 = 1000 \times a$$

$$a = \frac{300}{1000}$$

$$= 0.3\,\text{m/s}^2$$

Typical masses, accelerations and forces

You should know the approximate sizes of speeds, accelerations and forces involved in everyday transport. The table gives some examples for a car and a train.

	Mass	Acceleration from rest	Resultant force acting to accelerate from rest	Maximum speed
car	≈ 1500 kg	≈ 2 m/s²	≈ 3 000 N	≈ 30 m/s
train	≈ 200 × 10³ kg	≈ 0.2 m/s²	≈ 40 000 N	≈ 55 m/s

Maths note

You should be familiar with these mathematical symbols:
- proportional to: $\propto$
- approximately equal to: $\approx$

The effect of mass and force on acceleration

Figure 22.19 shows the apparatus you might use to investigate how acceleration of an object depends on the object's mass and the force applied to accelerate it.

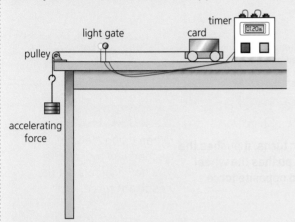

Figure 22.19

Measuring acceleration

The trolley is allowed to accelerate from rest.

The trolley's final speed is calculated:

$$v = \frac{\text{length of card}}{\text{time taken to pass through light gate}}$$

The trolley's acceleration is calculated:

$$a = \frac{\text{final speed, } v}{\text{time taken to reach the light gate}}$$

Changing the force

The mass of the trolley is kept constant.

The force is varied and accelerations are calculated for different forces to show that:

$a \propto F$

Doubling the force, for a fixed mass, doubles the acceleration.

Changing the mass

The force accelerating the trolley is kept constant.

The mass is varied and accelerations calculated for different masses to show that:

$a \propto \dfrac{1}{m}$

Doubling the mass, for a fixed force, halves the acceleration.

Newton's third law of motion

REVISED

Newton's third law states that whenever two objects interact, the forces they exert on each other are equal and opposite.

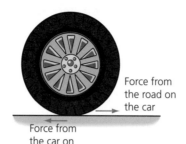

(a) If I push you with a force of 100 N, you push me back with a force of 100 N.

(b) When the wheel of a car turns, it pushes the road backwards. The road pushes the wheel forwards with an equal and opposite force.

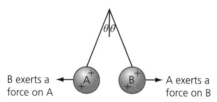

(c) A spacecraft orbiting the Earth is pulled downwards by the Earth's gravity. The spacecraft exerts an equal and opposite gravitational force on the Earth.

(d) Two balloons have been charged positively. They each experience a repulsive force from the other. These forces are of the same size, so each balloon (if of the same mass) is lifted through the same angle.

Figure 22.20

These are the features of Newton's third law pairs:
- they act on separate bodies
- they are always of the same type – for example, two gravitational forces or two contact forces
- they are of the same magnitude
- they act along the same line
- they act in opposite directions.

Now test yourself

TESTED

22 A feather falls to the ground at a constant speed. Which of the following is true?
   A   The feather's weight is slightly greater than the air resistance on the feather.
   B   The feather's weight is equal to the air resistance acting on it.
23 (a) A book is at rest on a table.
     (i)   Draw a diagram to show the two forces acting on the book – its weight, W, and the contact force from the table, R.
     (ii) Explain why the resultant force on the book is zero.
  (b) A student says:

     'Equal and opposite forces act on the book; this is an example of Newton's third law.'

     (i)   Explain why the student is wrong.
     (ii) What is the equal and opposite force to the weight of the book, as described by Newton's third law?

24 Explain why a racing car is designed to have as low a mass as possible.
25 A car has a mass of 1500 kg. It accelerates from 10 m/s to 18 m/s in 12 s.
   (a) Calculate the car's acceleration.
   (b) Calculate the resultant force acting on the car while it accelerates.
26 A boy blows up a balloon, and then releases it, so that air escapes from it. The balloon flies around the room. Use Newton's third law to explain the motion of the balloon.
27 The diagram shows a speed boat.
   (a) Explain why the boat is moving at a constant speed.

8000 N
forward force

8000 N
drag force

Figure 22.21

   (b) The engine speed is reduced so that the boat slows down. Use the information in the graph in Figure 22.22 to calculate the boat's deceleration over the region AB.

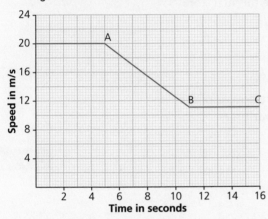

Figure 22.22

   (c) (i) The mass of the boat is 2500 kg. Calculate the resultant force on the boat while it decelerates.
       (ii) Calculate the forwards force on the boat due to the propellers, while the boat decelerates.
       (iii) State the size of the drag force acting on the boat over the region BC.

Answers online

Forces and braking

Driving: stopping distance

When a driver of a vehicle sees a hazard she reacts, applies the brakes and stops the vehicle.
- stopping distance = thinking distance + braking distance
- thinking distance = distance travelled while the driver reacts
- braking distance = distance travelled while the car brakes
- stopping distance = total distance travelled from when the driver first sees a hazard to the point where the car stops

Factors affecting reaction times

Reaction times vary from person to person; typical times vary from 0.2s to 0.9s.

Driver's reaction times are slower if they:
- have been drinking alcohol
- have been taking certain types of drugs
- are tired
- are distracted by using their mobile phone.

Exam tip

A common exam question is to ask you about the factors that affect reaction times and braking distance.

Factors affecting braking distance

- **Speed**. A large speed increases the braking distance.
- **Force**. A large braking force reduces the braking distance.
- **Mass**. Making the car more massive by carrying a large load increases the braking distance.
- **Weather**. In wet or icy conditions there is less friction between the road and tyres – braking distance increases.
- **Vehicle maintenance**. Worn brakes or worn tyres increase the braking distance.
- **Road condition**. Some road surfaces affect the friction on the tyres – a smooth or muddy road can increase the braking distance.

Braking and kinetic energy

The work done by the brakes reduces a vehicle's kinetic energy, causing the vehicle to stop. The kinetic energy is transferred to the thermal store in the brakes.

work done = kinetic energy transferred

$$Fs = \frac{1}{2}mv^2$$

This shows the braking distance is proportional to v^2 (and the mass of the car). If the speed doubles, the braking distance increases four times.

Now test yourself

28 Which of the following affects the thinking distance of a driver?
   an icy road   a drunk driver   worn brakes
29 Which of the following affect the braking distance of a car?
   a tired driver   a muddy road   the speed of the car
30 A car is travelling at 15 m/s. A driver has a reaction time of 0.4 s.
   (a) Calculate the thinking distance of the driver.
   (b) When the driver is tired, his reaction time increases to 0.6 s. Calculate the thinking distance now.

Answers online

Momentum

Momentum is defined by the equation:

momentum = mass × velocity

$$p = mv$$

Since velocity is a vector quantity, momentum is also a vector quantity.

momentum, p, in kilogram metres per second, kg m/s

mass, m, in kilograms, kg

velocity, v, in metres per second, m/s

Conservation of momentum

In a closed system, the total momentum before an event is equal to the total momentum after the event.

The word 'closed' means that no external forces act on the system. For example, if a moving car collides with a stationary car and they stick together, they share the momentum. Both cars move together, but at a lower speed than the moving car had before the collision.

31 Which of the following is the correct unit for momentum?

kg m s   kg m/s²   kg m/s

32 A car of mass 1250 kg travels at a speed of 20 m/s. Calculate the car's momentum.

Answers online

Summary

- Scalar quantities have magnitude only.
- Vector quantities have magnitude and directions.
- A force is a push or a pull; there are contact and non-contact forces.
  Force is measured in newtons, N.
- Weight is the pull of gravity on a mass.

  weight = mass × gravitational field strength

  $$W = mg$$

- A resultant force is the vector sum of a number of forces acting on a body.
- The work done by a force acting on an object is calculated by:

  work done = force × distance (moved along the line of action of the force)

  $$W = Fs$$

  Work done is measured in joules, J.
- The extension of an elastic object such as a spring is proportional to the force, provided the spring does not exceed its limit of proportionality.

  force = spring constant × extension

  $$F = ke$$

  k is measured in N/m.
- The work done in stretching (or compressing) a spring (up to the limit of proportionality) is calculated using:

  elastic potential energy $= \dfrac{1}{2} \times$ spring constant × (extension)²

  $$E_e = \dfrac{1}{2}ke^2$$

- distance travelled = speed × time
- When the speed changes during the motion, an average speed may be calculated:

  $$\text{average speed} = \dfrac{\text{distance travelled}}{\text{time}}$$

- The gradient of a distance–time graph is equal to the speed of the object travelling.

- $\text{acceleration} = \dfrac{\text{change of velocity}}{\text{time}}$

- The gradient of a velocity–time graph is equal to the acceleration.

H ● The 'area' under a velocity–time graph is equal to the distance travelled.
- The following equation applies to uniform motion:

  (final velocity)² − (initial velocity)² = 2 × acceleration × distance

  $$v^2 - u^2 = 2as$$

- Acceleration due to Earth's gravity: $g = 9.8\,\text{m/s}^2$.
- A falling object reaches a terminal velocity when the resistive forces balance the object's weight.
- Newton's first law states that when the resultant force on an object is zero:
  the object remains at rest or moves with a constant speed in a straight line.
- Newton's second law states that:

  resultant force = mass × acceleration

  $$F = ma$$

- Newton's third law states that:
  whenever two objects interact, the forces they exert on each other are equal and opposite.

- stopping distance = thinking distance + braking distance

- Reaction times are slowed if a driver is tired, consumes alcohol or some drugs, or is distracted.
- Braking distances are larger at high speeds, when carrying a massive load, in wet or icy conditions, or if the tyres or brakes of a vehicle are worn.

- momentum = mass × velocity

H ● In a closed system, the total momentum of the system is conserved.

Exam practice

1 Which one of the following is a vector quantity? [1]
  force   speed   energy
2 How many newtons are there in a kilonewton? [1]
  10   1000   1 000 000
3 Which of the following is the correct unit for energy? [1]
  newton   joule   watt
4 An astronaut in his spacesuit has a mass of 120 kg on the Earth, where the gravitational
  field strength is 9.8 N/kg.
  (a) Calculate his weight. [2]
  (b) He climbs up a ladder 4.5 m high into the spacecraft.
      Calculate the work done by his legs as he climbs. [3]
  (c) Explain why the astronaut has a smaller weight when he lands on Mars. [2]

Required practical 18

5 A student carried out an investigation into the stretching of a spring A. He makes a hypothesis that
  the extension of the spring will be proportional to the weight on it.
  Figure 22.23 shows the spring in (i) its unstretched state and (ii) when it has been stretched by a
  load.

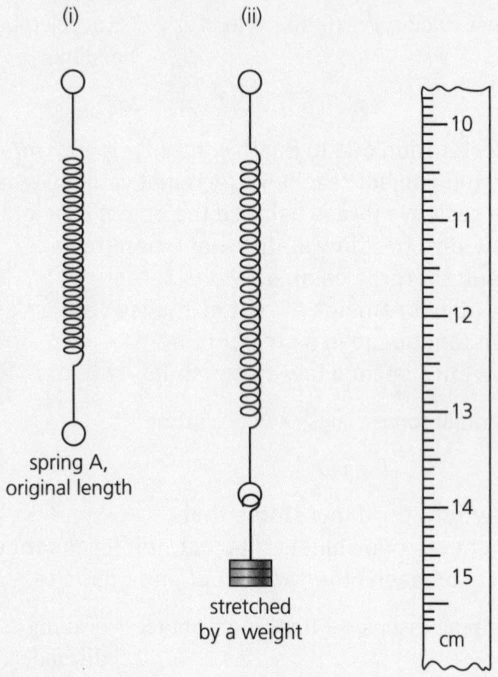

Figure 22.23

  (a) Use Figure 22.23 to calculate the spring's extension in this case. [2]
  (b) The student uses weights up to 5 N to stretch spring A. The results are shown in Figure 22.24.

→

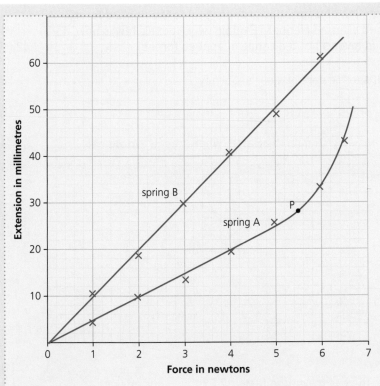

Figure 22.24

  (i) Explain whether this graph supports the student's hypothesis. [2]
  (ii) Not all the points plotted by the student lie on the straight line. Explain what type of error caused this and how you would attempt to reduce this type of error. [2]
  (iii) The student increases the load to 6.5 N. Explain what happens to the spring now. [1]
  (iv) Calculate the spring constant for spring A. Express your answer in N/m. [3]
(c) The student repeats his experiment for a second spring B (Figure 22.25). His results are also shown in Figure 22.24.
  (i) State which spring is stiffer. Explain your answer. [2]
  (ii) Use the graph to predict the extension of both springs together when a load of 4 N is used to extend them.

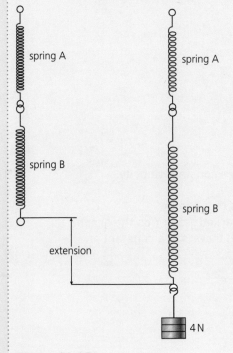

Figure 22.25

(d) Hooke's law states that 'the extension of a spring is proportional to the applied force'. This is an example of a law named after a scientist. Explain why Hooke's law is now accepted to be true for all springs (over a limited range of applied forces). [2]

6 Figure 22.26 shows a distance–time graph for part of a car's journey.

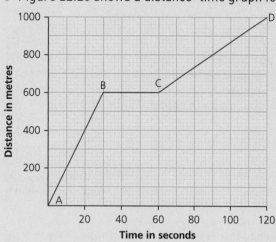

Figure 22.26

   (a) How long was the car stopped for? [1]
   (b) Over which part of the journey was the car travelling fastest? Explain your answer. [2]
   (c) Calculate the average speed for the whole journey from A to D. [2]

7 A small ball with a weight of 0.5 N is allowed to fall from rest. It accelerates until it reaches a constant velocity.

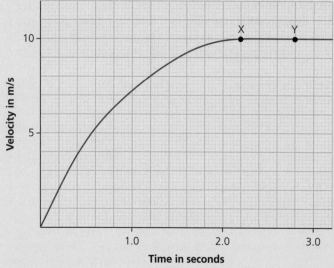

Figure 22.27

   (a) Use the graph to determine the time it took the ball to reach a constant velocity. [1]
   (b) Explain the shape of the graph. [3]
   (c) Calculate the acceleration of the ball in the first 0.4 seconds that it falls. [3]
   (d) Calculate the distance the ball travels between the times marked X and Y on the graph. [2]
   (e) State the size of the resultant force acting on the ball when it travels at a constant velocity. Explain your answer. [2]

→

8 A car driver is travelling at 25 m/s on a motorway when she sees a hazard ahead. The car travels a further 15 m, before the driver presses the brake pedal.
  (a) Calculate the driver's reaction time. [2]
  (b) State two factors that could affect her reaction time. [2]
  (c) The car comes to a halt after a distance of 60 m. Use the equation

  $v^2 - u^2 = 2as$

  to calculate the car's deceleration. [3]
  (d) State two factors which could affect the car's braking distance. [2]
  (e) On another occasion the car is travelling at 12.5 m/s. State whether the braking distance will be more than, equal to, or less than 30 m. Explain your answer. [2]
9 A skydiver has a mass of 100 kg including his parachute. Gravitational field strength $g = 9.8$ N/kg.
  (a) Calculate the skydiver's weight. [2]
  (b) Before he opens his parachute he falls with a constant velocity of 50 m/s. State what the resultant force acting on him is. [1]
  (c) The skydiver opens his parachute, which causes the drag force on him to increase to 1480 N.
    (i) State the resultant force acting on a skydiver now. [1]
    (ii) Calculate the deceleration of the skydiver when he opens the parachute. [2]
10 Use Newton's laws of motion to explain each of the following:
  (a) You can throw a tennis ball of mass 60 g much further than you can put a shot of mass 4 kg. [3]
  (b) Two ice skaters, Paul and Jane, stand next to each other on the ice. Paul gives Jane a push. Both ice skaters begin to move. Explain why. [3]
  (c) A parcel is placed on the seat of a car. When the car brakes suddenly the parcel falls onto the floor. Explain why. [3]
11 Describe an experiment to measure the acceleration of an object in a laboratory. Describe the apparatus you would use and explain what measurements you would take to calculate the acceleration. Explain what action you would take to ensure the experiment is safe and does not damage any apparatus. [6]

Answers and quick quiz 22 online

ONLINE

23 Waves

All waves carry energy and information. Sound waves allow us to hear, and electromagnetic waves let us see and enable communications via phone, internet, television and radio.

We study waves on water and on ropes or springs, so that we can apply that understanding to sound and electromagnetic waves, which we cannot see.

Waves in air, fluids and solids

When a stone lands in a pond, you see water ripples spreading out. The circular ripples give us the information about where the stone landed (if you did not see it). The energy is transmitted through the water, but the water itself does not move outwards. These ripples are examples of waves.

Figure 23.1

Transverse waves

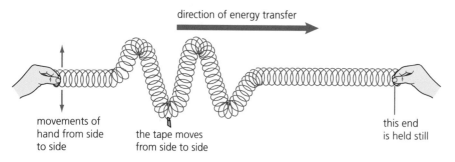

direction of energy transfer

movements of hand from side to side

the tape moves from side to side

this end is held still

Figure 23.2 The coloured tape shows that the slinky coils move from side to side, like the hand.

Figure 23.2 shows a **transverse wave**. The student moves the slinky from side to side, and the energy is transmitted along the spring. The water waves in Figure 23.1 are also examples of transverse waves. Electromagnetic waves (including light) are transverse waves.

> A **transverse wave** is one in which the vibrations causing a wave are at right angles to the direction of energy transfer.

Longitudinal waves

Figure 23.3 shows a **longitudinal wave**. Energy is transmitted along the slinky when it is pushed backwards and forwards. In some places coils are pushed together – these are compressions (C); in other places the coils are pulled apart, these are rarefactions (R). Sound waves are examples of longitudinal waves.

> A **longitudinal wave** is one in which the vibration causing the wave is parallel to the direction of energy transfer.

direction of the vibration

this end is held still

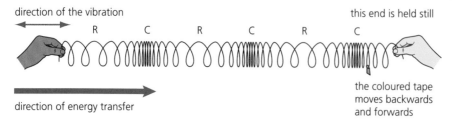

R   C   R   C   R   C

direction of energy transfer

the coloured tape moves backwards and forwards

Figure 23.3 The coloured tape shows that the slinky coils move backwards and forwards like the hand.

Answers and quick quizzes at **www.hoddereducation.co.uk/myrevisionnotes**

Now test yourself

1 (a) Use diagrams to illustrate the nature of:
   (i) transverse waves
   (ii) longitudinal waves.
 (b) Give an example of each type of wave mentioned in part (a).
2 Describe how you would use a slinky to show that waves carry both energy and information.
3 Figure 23.4 shows a water wave approaching a ball. Copy the diagram and show the movement of the ball as the wave reaches it.

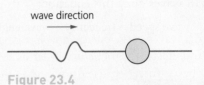

wave direction

Figure 23.4

Answers online

Properties of waves

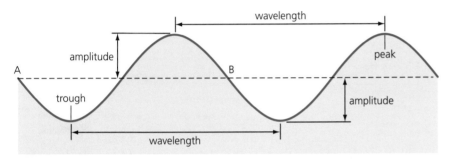

wavelength
amplitude
peak
A
B
trough
amplitude
wavelength

Figure 23.5

- The **amplitude**, A, is the distance from a wave peak to the middle, or from a wave trough to the middle.
- The **wavelength**, λ, is the distance between two adjacent peaks or two adjacent troughs.
- The **frequency**, f, of a wave is the number of complete waves produced per second. The unit of frequency is the **hertz (Hz)**. Large frequencies are measured in kilohertz, kHz (10^3 Hz) or megahertz, MHz (10^6 Hz).
- The **period**, T, of a wave is the time taken to produce one wave.

The frequency and time period are linked by the equation:

$$\text{period} = \frac{1}{\text{frequency}}$$

$$T = \frac{1}{f}$$

period, T, in seconds, s
frequency, f, in hertz, Hz

> **Amplitude** is the height of the wave from the middle (undisturbed) position of the string or water.
>
> **Wavelength** is the distance from one point on a wave to the equivalent point on the next wave.
>
> **Frequency** is the number of waves passing a point each second.
>
> **Hertz (Hz)** is the unit of frequency.
>
> **Period** is the time taken to produce one wave – or the time taken for one wave to pass a point.

Example

A wave has a period of 0.14 s. Calculate its frequency.

Answer

$$T = \frac{1}{f}$$

$$0.14 = \frac{1}{f}$$

$$f = 7.1\,\text{Hz}$$

Wave speed and the wave equation

The wave speed is the speed at which energy is transferred (or the wave moves) through a medium.

All waves obey this equation:

wave speed = frequency × wavelength

$$v = f\lambda$$

wave speed, v, in metres per second, m/s
frequency, f, in hertz, Hz
wavelength, λ, in metres, m

Example

Sound travels at 330 m/s in air. Calculate the wavelength of a sound wave having a frequency of 900 Hz.

Answer

$$v = f\lambda$$

$$330 = 900 \times \lambda$$

$$\lambda = \frac{330}{900}$$

$$= 0.37\,m$$

Now test yourself

4 Figure 23.6 shows waves travelling at the same speed on two ropes. Use the correct terms to describe the differences between the waves on rope A and rope B.

rope A   rope B

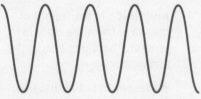 

Figure 23.6

5 (a) Two waves have periods of:
   (i) 0.02 s
   (ii) 0.001 s
   Calculate their frequencies.
(b) Two waves have frequencies of:
   (i) 10^9 Hz (10^9 Hz = 1 gigahertz = 1 GHz)
   (ii) 2 MHz
   Calculate their periods.

6 A stone is thrown into a pond and waves spread out. The speed of the waves is 1.2 m/s and their wavelength 20 cm.
   Calculate the frequency of the waves.

7 Figure 23.7 shows a graph of a wave on a slinky: the *x*-axis shows the distance along the slinky, and the *y*-axis shows the displacement of the slinky from its undisplaced position.

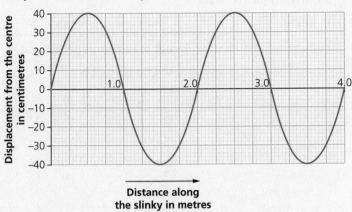

Figure 23.7

(a) What type of wave is moving along the slinky?
(b) (i) State the wave's amplitude.
    (ii) State the wave's wavelength.
(c) The teacher's hand, that produces the waves, moves backwards and forwards from +40 cm to −40 cm and back to +40 cm twice per second.
    (i) State the frequency of the waves.
    (ii) State the time period of the waves.
(d) Calculate the speed of the waves on the slinky.

Answers online

Measuring wave speeds

The speed of sound

REVISED

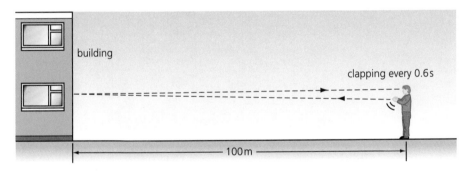

Figure 23.8

- A student claps his hands, and he claps again as he hears the echo from the building.
- His friend finds that he does 10 claps in 6 seconds.
- He calculates the speed of sound.

$$v = \frac{d}{t}$$

$$= \frac{200\,\text{m}}{0.6\,\text{s}}$$

$$= 330\,\text{m/s}$$

> **Typical mistake**
>
> In measuring the speed of sound, remember that the sound has to go to the building **and back**. So the distance travelled here is 200 m not 100 m.

Required practical 20

(a) Investigating waves in a ripple tank

● A ripple tank is set up and the wave generator is attached to a variable frequency supply; the frequency is 12 Hz.
● The pattern seen in Figure 23.9 can be photographed using a mobile phone.
● Now we can work out the wave speed. Since 8 wave crests can be seen in the ripple tank length of 32 cm:

$$\text{wavelength} = \frac{32\ \text{cm}}{8}$$

$$= 4\ \text{cm}$$

$$v = f\lambda$$

$$= 12 \times 4$$

$$= 48\ \text{cm/s}$$

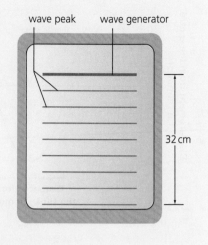

Figure 23.9

(b) Waves in a stretched spring

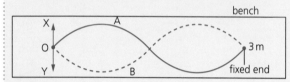

Figure 23.10

● In Figure 23.10 a slinky is stretched along a bench. It is 3.0 m long.
● A student moves one end from X to Y and back to X, so that one wavelength fits into the spring length. Sometimes the shape looks like A, then we see pattern B.
● Another student measures the time for 10 complete oscillations X → Y → X. He finds 10 oscillations take 6.7 seconds.
● Now we can work out the wave speed.

$$\text{wavelength}, \lambda = 3.0\ \text{m}$$

$$\text{period}, T = \frac{6.7}{10}$$

$$= 0.67\ \text{s}$$

$$\text{frequency}, f = \frac{1}{T}$$

$$= \frac{1}{0.67}$$

$$= 1.5\ \text{Hz}$$

$$\text{speed} = \text{frequency} \times \text{wavelength}$$

$$= 1.5 \times 3.0$$

$$= 4.5\ \text{m/s}$$

We can check our result by a second experiment shown in Figure 23.11.

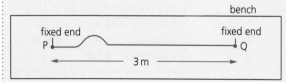

Figure 23.11

Answers and quick quizzes at www.hoddereducation.co.uk/myrevisionnotesdownloads

- Both ends of the slinky are fixed, and a wave pulse is produced by pulling the slinky to one side and letting it go. The pulse takes 2.7 s to go up and down the slinky twice, a distance of 12 m.
- Now we can calculate the speed as the distance travelled is 12.0 m in 2.7 s:

$$v = \frac{d}{t}$$

$$= \frac{12.0}{2.7}$$

$$= 4.4 \, \text{m/s}$$

Now test yourself

8 Describe experiments to measure:
(a) the speed of sound
(b) the speed of a wave along a stretched spring.
In each case, state the measurements you would take and explain how you would use the measurement to calculate the wave speed.

Answers online

Electromagnetic waves

Electromagnetic waves are transverse waves that transfer energy from the source to an absorber.

Electromagnetic waves form a continuous spectrum, with wavelength varying from 10^{-12} m to over 1 km. Electromagnetic waves are grouped by wavelength. These groups are: radio, microwave, infrared, visible light, ultraviolet, x-rays and gamma rays.

Electromagnetic waves all have these properties.
- They are transverse waves.
- They transfer energy from one place to another.
- They obey the equation $v = f\lambda$.
- They travel through a vacuum.
- They all travel at the same speed in a vacuum – 300 000 000 m/s (3×10^8 m/s).

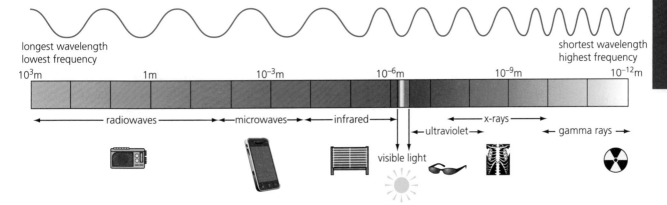

Figure 23.12

Properties of electromagnetic waves 1

Different substances absorb, transmit, refract or reflect electromagnetic waves in ways that vary with wavelength.

- We see different colours – a green shirt reflects green light, but absorbs all other colours.
- A polished metal surface reflects wavelengths of electromagnetic waves from radio waves to ultraviolet, but x-rays and gamma rays are transmitted by thin metal plates.
- Food is cooked in a microwave oven. The wavelength of microwaves is carefully chosen so that they are absorbed by water.

Refraction

Figure 23.13 shows the **refraction** of a light ray as it passes through a glass block.

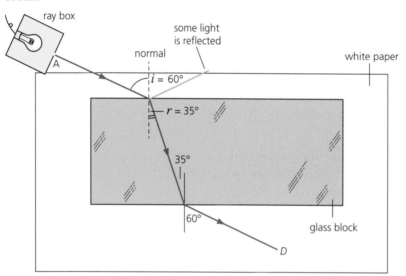

Figure 23.13

> When a wave is transmitted from one medium to another, the transmitted wave changes direction. This is called **refraction**. All types of waves, including light and sound, refract when they travel from one medium to another.

- Light is refracted when it enters and leaves glass (or water) because the speed of light is greater in air than it is in glass (or water).
- Light bends towards the normal when it enters glass, and away from the normal when it travels from glass into air.

All waves show refraction.

- Water waves are refracted when they travel from deep water to shallow water, as shown in Figure 23.14.

Here the waves:
- slow down
- become shorter in wavelength
- change direction.

Exam tip

Note that the waves do not refract (change direction) when they travel parallel to the normal.

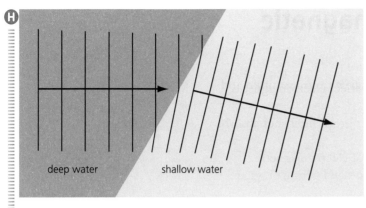

Figure 23.14

Now test yourself

9 The speed of electromagnetic waves is 3×10^8 m/s.
  (a) Red light has a wavelength of 6.5×10^{-7} m. Calculate the frequency of red light.
  (b) A radio wave has a frequency of 2 MHz. Calculate its wavelength.

10 Draw diagrams to show how the water waves are refracted in each of the following diagrams.

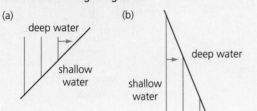

Figure 23.15

> **Exam tip**
>
> It is easier to use standard form to solve problems with very large or very small numbers.

11 Draw diagrams to show how light is refracted in each of the following cases.

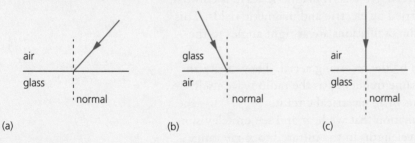

Figure 23.16

Answers online

Properties of electromagnetic waves 2

Figure 23.17 shows how radio waves and microwaves are transmitted and received.

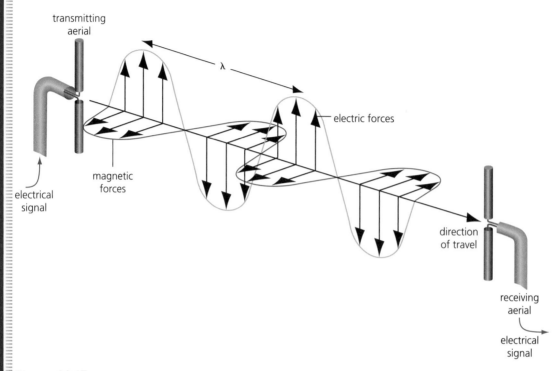

transmitting aerial

λ

electric forces

electrical signal

magnetic forces

direction of travel

receiving aerial

electrical signal

Figure 23.17

- A high frequency alternating potential difference causes electrons to oscillate in the transmitting aerial. An electromagnetic wave is emitted.
- The energy in the wave is carried by electric and magnetic fields. This is a transverse wave because the oscillations are at right angles to the energy transfer.
- The radio waves are absorbed by the receiving aerial. They create an alternating current with the same frequency as the radio wave itself. So radio waves induce oscillations in an electrical circuit.
- Radio waves also carry information that we hear and see on televisions.
- Electromagnetic waves of wavelengths in the infrared to x-ray range are produced by changes in atoms. Gamma rays originate from changes in the nucleus of an atom.

Hazards of radiation

REVISED

Ultraviolet waves, x-rays and gamma rays can be hazardous to humans.
- Ultraviolet waves can cause skin to age prematurely and increase the risk of skin cancer. Ultraviolet waves from the Sun cause sun tanning.
- X-rays and gamma rays are ionising radiations that can cause the mutation of genes and cancer. Radiation doses are measured in Sieverts.
- Sunburn is caused by infrared radiation.

Uses and applications of electromagnetic waves

Electromagnetic waves have many practical applications. For example:
- radiowaves – television and radio
- microwaves – satellite communications, cooking food
- infrared – electrical heaters, cooking food, infrared cameras
- visible light – sight, fibre optic communications
- ultraviolet – energy efficient lamps
- X-rays and gamma rays – medical imaging and treatments.

(H)
- Radio waves are used to transmit radio and television signals. Long wavelengths are suitable for transmissions around the Earth's surface.
- Microwaves are used for satellite communications and cooking. A narrow beam of microwaves is suitable for directing towards satellites. Some microwaves have the correct wavelength to be absorbed by water, thus allowing food to cook.
- Infrared radiation produces a heating effect, so we use infrared waves in electric heaters and in ovens for cooking. Warm objects also emit infrared waves; so an infrared camera allows us to see things at night. Infrared waves are also used in remote controls.
- We use visible light all the time to see. Light (or infrared) is also used in fibre optic communications. Fibre optic links are widely used in telecommunications.
- Some substances can absorb ultraviolet radiation and then emit the energy as visible light. This is called fluorescence. This principle is used in some energy efficient lamps. Fluorescence also has applications in crime solving. Possessions can be marked with an invisible fluorescent spray, which can be seen in ultraviolet light.
- X-rays (and gamma rays) can penetrate our bodies and can be used in medical imaging.
- Gamma rays (and x-rays) can be used in medical treatment. These radiations can be hazardous to our bodies, but they can also be used effectively to destroy cancerous tissue.

Now test yourself

TESTED

12 (a) Name two parts of the electromagnetic spectrum that are hazardous to humans.
   (b) For each of your choices explain what the hazards are.
13 Give and explain a use for:
   (a) microwaves
   (b) infrared waves.
14 List the seven types of wave in the electromagnetic spectrum starting with the waves with the lowest frequency.

Answers online

Required practical 21

Investigating the emission of infrared radiation from different surfaces

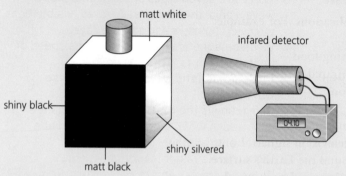

Figure 23.18

The apparatus shows a 'Leslie Cube'. It is filled with hot water. The infrared detector is placed close to each face to detect how much radiation is emitted from each surface. We make it a fair test by keeping the detector the same distance from each face.

This investigation shows us that:

● dull black surfaces are good emitters of infrared radiation
● shiny or white surfaces are poor emitters of infrared radiation.

We can also investigate which type of surfaces are good absorbers of radiation.

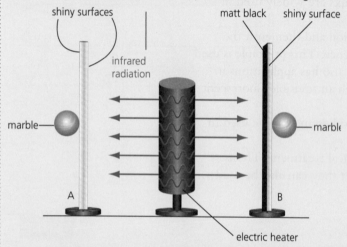

Figure 23.19

In Figure 23.19 two marbles are stuck with wax onto two metal sheets. One sheet (B) has a dull black surface facing the electric heater; the other sheet (A) has a shiny metallic surface facing the heater.

The marble falls quickly from side B, and remains on side A.

● Dull black surfaces are good absorbers of infrared radiation.
● Shiny or white surfaces are poor absorbers of infrared radiation.

Now test yourself

TESTED ☐

15 In the experiment with the marbles (Figure 23.19), explain what measures must be taken to make it a fair test.
16 Below are four types of surface.
   dull black   shiny black   dull white   shiny metallic
   Organise the list in the order of:
   (a) the best emitters of infrared radiation
   (b) the best absorbers of infrared radiation.

Answers online

Summary

- Waves transfer energy and information.
- In a transverse wave the vibrations of the wave are at right angles to the direction of energy transfer.
- In a longitudinal wave the vibrations are parallel to the direction of energy transfer.
- Amplitude, A, is the height of a wave measured from its undisturbed position.
- Wavelength, λ, is the distance between two adjacent peaks (or troughs).
- Frequency, f, is the number of waves produced per second.
- Period, T, is the time taken to produce one wave.
- $f = \dfrac{1}{T}$

- $v = f\lambda$
- There are seven types of electromagnetic wave: radio, microwaves, infrared, visible light, ultraviolet, x-rays and gamma rays.
- Electromagnetic waves travel at the speed of light: 3×10^8 m/s (in a vacuum).
- Electromagnetic waves are transverse waves.
- Refraction is the name given to the change of direction when a wave travels from one material to another.
- Light travels faster in air than in glass.
- When a light ray enters glass from air, it bends towards the normal. The ray bends away from the normal when it travels from glass to air.

Exam practice

1 Figure 23.20 shows transverse waves on a piece of thick string. A person produces the waves by holding the string and moving it from side to side. The end A travels through 5 complete movements (from side to side and back) in 2 seconds.

   (a) Explain what is meant by a transverse wave. [2]

   (b) Use the diagram to calculate:

      (i) the amplitude of the waves [1]

      (ii) the wavelength of the waves. [1]

   (c) Calculate:

      (i) the frequency of the waves [1]

      (ii) the period of the waves. [1]

   (d) Calculate the speed of the waves. [3]

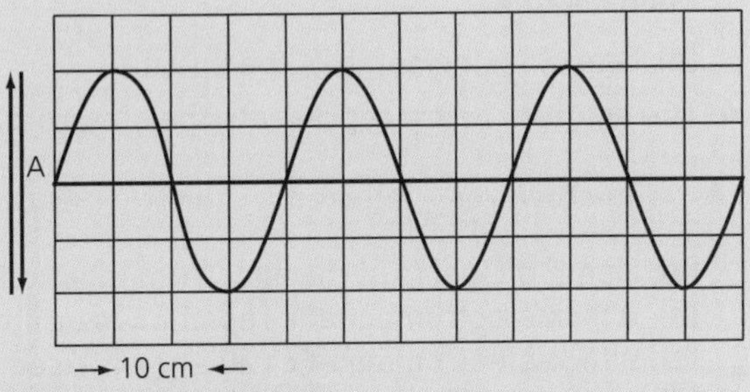

A

— 10 cm —

Figure 23.20

2 Figure 23.21 shows water waves in a ripple tank. The waves are travelling in deep water. The waves cross a boundary into shallow water, where they travel more slowly.

   Copy the diagram and show what happens to the waves when they travel in the shallow water. [3]

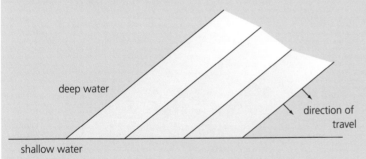

deep water

direction of travel

shallow water

Figure 23.21

3 Figure 23.22 shows how the depth of the oil in a storage tank is measured using sound of frequency 15 kHz. The sound is emitted and received by a transducer at the bottom of the tank. Sound is reflected off the surface of the oil. The time interval between the transmitted and reflected waves is displayed on the oscilloscope.

(a) State the range of frequencies that a human can hear. [1]

(b) The speed of sound in the oil is 1200 m/s. Calculate the wavelength of the waves, using the information above. [3]

(c) Use the information on the oscilloscope trace to show that the time taken to travel from the transducer to the surface and back is 8 ms. [1]

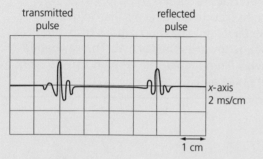

transmitted pulse   reflected pulse

x-axis
2 ms/cm

1 cm

Figure 23.23

oil storage tank

transducer that sends and receives high frequency sound waves

Figure 23.22

(d) Calculate the depth of the liquid. [3]

4 The diagram below represents the electromagnetic spectrum

A	microwaves	infrared	light	B	X-rays	gamma rays

(a) Name the parts of the spectrum labelled A and B. [2]
(b) Which electromagnetic radiation is emitted by a hot oven? [1]
(c) Which electromagnetic radiation is used for communication with satellites? [1]
(d) An X-ray has a wavelength of 1.2×10^{-10} m.
    Calculate the frequency of the wave. The speed of light is 3.0×10^8 m/s. [3]

Answers and quick quiz 23 online

ONLINE

24 Magnetism and electromagnetism

Permanent and induced magnetism, magnetic forces and fields

Poles

The poles of a magnet are the places where the **magnetic** forces are strongest. When two magnets are brought close to each other they exert a force on each other.

- Two like poles repel each other.
- Two unlike poles attract each other.

Magnetic forces of attraction and repulsion are examples of non-contact forces.

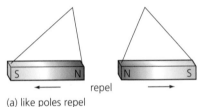

(a) like poles repel   (b) unlike poles attract

Figure 24.1 Two like poles repel, two unlike poles attract.

There are two types of pole, north pole and south pole. A north pole is short for a **north-seeking pole**, and south pole is short for a **south-seeking pole**. If a magnet is suspended, it points along the north–south direction.

> **Magnetic** materials are attracted by a magnet.
>
> **A north-seeking pole** of a magnet points north.
>
> **A south-seeking pole** of a magnet points south.

Permanent and induced magnets

A **permanent magnet** produces its own magnetic field.

A permanent magnet always has a north and a south pole. If you have two permanent magnets, you will be able to show that they can repel each other, as well as attract.

An **induced magnet** is a material that becomes magnetic when it is placed in a magnetic field. It loses its magnetism when it is removed from a magnetic field. An induced magnet is always attracted to a permanent magnet, because the induced magnet is magnetised in the direction of the magnetic field.

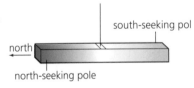

Figure 24.2 The north-seeking pole points towards north, and the south-seeking pole points towards south.

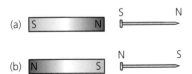

Figure 24.3 The nail becomes magnetised in the direction of the permanent magnet's field. The nail is always attracted to the permanent magnet.

> A **permanent magnet** produces its own magnetic field.
>
> An **induced magnet** becomes a magnet when it is placed in a magnetic field.

Magnetic fields

The region around a magnet, where a force acts on another magnet or a magnetic material, is called a magnetic field.

The strength of the magnetic field depends on the distance from the magnet. The field is strongest near the poles of the magnet. We use magnetic field lines to represent a magnetic field. Magnetic field lines always start at a north pole and finish at a south pole. When the lines are close together, the field is strong. The further apart the lines are, the weaker the field is.

The direction of a magnetic field can be found using a small plotting compass. The compass needle always points along the direction of the field, as shown in Figure 24.4

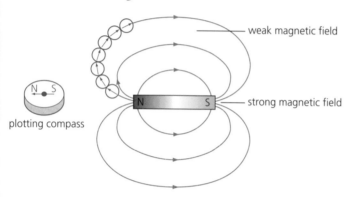

weak magnetic field

strong magnetic field

plotting compass

Figure 24.4

The Earth's magnetic field

We use a compass to help us navigate. The Earth has a magnetic field. The north (seeking) pole of the magnet points towards magnetic north. Figure 24.5 shows the shape of the Earth's field. The north pole of the compass is attracted towards a south (seeking) pole at magnetic north.

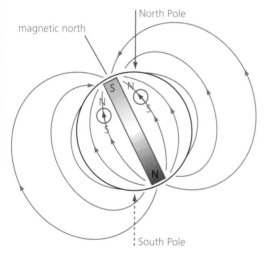

Figure 24.5 The shape of the Earth's magnetic field.

Now test yourself

1 Name four magnetic materials.
2 How many types of magnetic pole are there? State the rule about the attraction and repulsion of magnetic poles.
3 (a) Explain what is meant by a *permanent magnet*.
  (b) Explain how you can show a magnet is a permanent magnet.
4 Draw the shape and direction of a magnetic field around the bar magnet.
5 What type of pole is there at the magnetic north?
6 In Figure 24.6, three steel paper clips are attached to a magnet.
  (a) The size of the magnetic force is greater than which other force acting on each paper clip?
  (b) (i) What type of magnets are the paper clips?
      (ii) Draw a diagram to show the magnetic poles on each paper clip.
  (c) Explain why the paper clips could be suspended from the south pole of the magnet.
7 Explain why the steel pins repel each other in Figure 24.7.

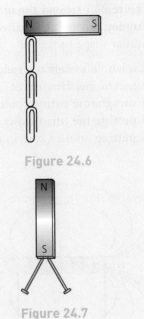

Figure 24.6

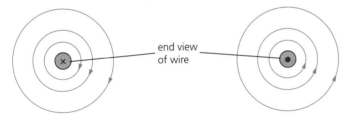

Answers online

Figure 24.7

The motor effect

Electromagnetism

When a current flows through a conducting wire, a magnetic field is produced around the wire. The strength of the field:
● depends on the size of the current through the wire
● is weaker further away from the wire.

end view of wire

Figure 24.8

Figure 24.8 shows the pattern of magnetic field lines surrounding a wire. When the current flows into the paper (shown as ⊗) the field lines are clockwise. When the current flows out of the paper (shown as ⊙) the field lines are anticlockwise.

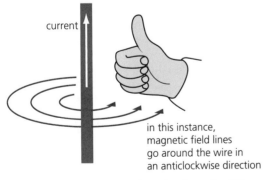

current

in this instance, magnetic field lines go around the wire in an anticlockwise direction

Figure 24.9 The right-hand grip rule. When you put your thumb along the direction of the current in a wire, your fingers point in the direction of the magnetic field around the wire.

The field of a solenoid

Figure 24.10 shows the magnetic field that is produced by a current flowing through a long coil of wire (a **solenoid**). The magnetic field has a similar shape to that of a bar magnet.

> A **solenoid** is a long coil of wire.

A solenoid's magnetic field can be increased by:
● increasing the current
● using more turns of wire
● putting the turns closer together
● putting an iron core in the middle of the solenoid.

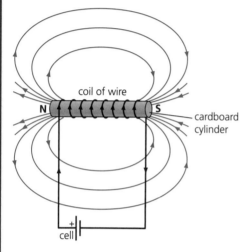

Figure 24.10 The magnetic field around a solenoid has a similar shape to that of a bar magnet.

Figure 24.11

Electromagnets

Figure 24.11 shows an electromagnet in action. The iron core is an induced magnet, which is magnetised when a current flows through the coils. When the current is switched off, the magnet loses its magnetism and the iron filings fall off.

Now test yourself

8 List four ways of increasing the strength of the magnetic field produced by a solenoid.
9 Figure 24.12 shows a wire placed vertically with a current flowing into the paper. Copy the diagram, and add lines to show the direction and strength of the magnetic field around the wire.
10 Explain how to use the right-hand grip rule, to find the direction of the magnetic field around a wire that carries a current.
11 Sketch the shape of the magnetic field around a solenoid with a current flowing through it.
12 Figure 24.13 shows a solenoid wrapped on a hollow tube.
    (a) State the directions of each of the compass needles 1–6, e.g. pointing to the right, left, up or down.
    (b) Which end of the solenoid acts as a north pole?
    (c) State what happens to the compass needles when the current is reversed.

Figure 24.12

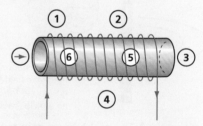

Figure 24.13

Answers online

Answers and quick quizzes at **www.hoddereducation.co.uk/myrevisionnotesdownloads**

⊕ Fleming's left-hand rule

When a conductor carrying a current is placed in a magnetic field, the magnet producing the field and the conductor exert a force on each other. This is called the motor effect.

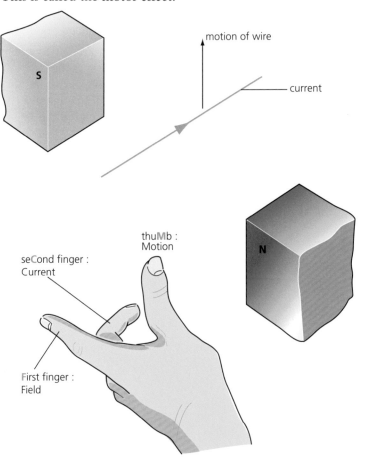

Figure 24.14

The size of the force on the wire depends on:
● the strength of the magnetic field from the magnet
● the size of the current
● the length of the wire between the poles of the magnet.

The left-hand rule allows us to predict the direction of the force on the wire. You arrange your thumb, first finger and second finger so that they are at right angles to each other, as shown in Figure 24.14.
● The first finger points in the direction of the magnetic field – north to south.
● The second finger points in the direction of the current.
● Then the thumb points along the direction of the force that causes the wire to move.

This rule works when the current and field are at right angles to each other. When the field and current are parallel to each other, the force on the wire is zero.

You can use the left-hand rule to predict that the direction of the force is reversed if:
● the magnetic field is reversed
● the current direction is reversed.

ⒽMagnetic flux density

We represent magnetic fields by drawing lines that show the direction of a force on a north pole. These field lines are formally known as **lines of magnetic flux**.

Figure 24.15 shows the lines of flux between two pairs of magnets. The magnets in Figure 24.15 (b) are stronger than the magnets in Figure 24.15 (a), so they exert a stronger force on a magnetic material.

We show stronger magnets by drawing more lines of flux in a given area.

The strength of the magnetic force is determined by the **flux density, B**.

> **Flux density**, B, is the number of lines of flux in a given area. (Flux density is sometimes called the B-field.)

Calculating the force

The force on a wire of length, L, carrying a current, I, at right angles to a magnetic field is given by the equation:

force = magnetic flux density × current × length

> force, F, in newtons, N
>
> magnetic flux density, B, in tesla, T
>
> current, I, in amperes (amps), A
>
> length, L, in metres, m

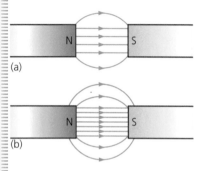

(a)

(b)

Figure 24.15

$$F = BIL$$

> ### Example
>
> Calculate the force on the wire in Figure 24.16. In which direction does the force act?
>
>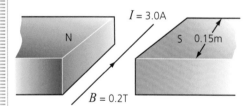
>
> Figure 24.16
>
> Answer
>
> $F = BIL$
>
> $= 0.2 \times 3.0 \times 0.15$
>
> $= 0.09\,\text{N} \approx 0.1\,\text{N}$
>
> Using the left-hand rule, the force acts downwards.

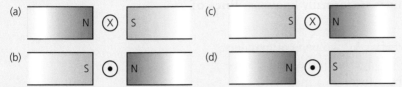

13 Predict the direction of the force on the wire in each of the following cases.

(a) N ⊗ S   (c) S ⊗ N

(b) S ⊙ N   (d) N ⊙ S

Figure 24.17

14 State two ways of increasing the force on a wire carrying a current in a magnetic field.
15 State the unit of magnetic flux density.
16 A wire of length 0.05 m is placed at right angles to a region of magnetic flux density 0.18 T.
   A current of 1.3 A flows through the wire.
   Calculate the force acting on the wire when
   (a) the field and current are at right angles to each other
   (b) the field and current are parallel.
17 In Figure 24.14, a magnetic force acts on the wire that carries a current. Newton's third law states that: whenever two objects interact, the forces they exert on each other are equal and opposite. On which object or objects does the wire exert a force?

Answers online

Electric motors

A coil carrying a current in a magnetic field tends to rotate. This is the basis of an electric motor.

In Figure 24.18 an upwards force acts on the wire AB, and a downwards force acts on the wire CD. These forces cause the coil to rotate clockwise. (Use the left–hand rule to check the directions of the force.) The coil stops rotating when it is vertical, because the two magnetic forces lie on the same vertical line.

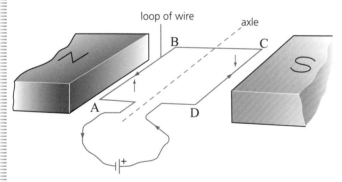

Figure 24.18 When the current flows the coil tends to rotate.

Figure 24.19 shows the design of a motor that allows the coil to keep turning. A split-ring commutator allows the direction of the current in the coil to reverse, once the coil passes the vertical position. Now the forces act to keep the coil rotating in a clockwise direction.

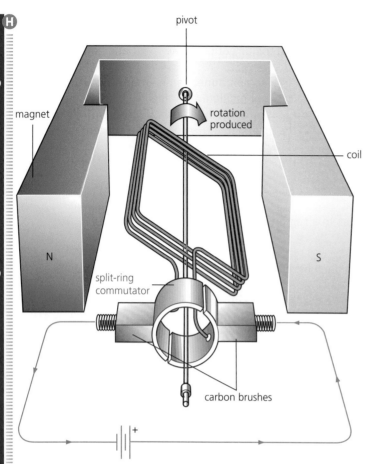

Figure 24.19 **A simple motor.**

A **split-ring commutator** allows the direction of current to reverse in a motor coil. This keeps the motor rotating in the same direction.

Now test yourself

18 Refer to Figure 24.18.
    (a) State the size of the force on the wire BC.
    (b) Explain why the coil will rotate and stop in a vertical position. You can draw diagrams to help your explanation.
19 Refer to Figure 24.19.
    (a) State three factors that affect the size of the forces that turn the coil.
    (b) Explain the function of the split-ring commutator.

Answers online

Summary

● Magnets have two poles, north and south.
● Like poles repel, unlike poles attract.
● Magnetic materials include: iron, steel, cobalt and nickel.

● A permanent magnet produces its own magnetic field. Two permanent magnets can attract or repel each other.
● An induced magnet becomes magnetic in a magnetic field. Induced magnets are always attracted towards permanent magnets.

→

- Magnetic fields: you should be able to sketch the shapes of the field close to a bar magnet, a wire carrying a current and a solenoid.

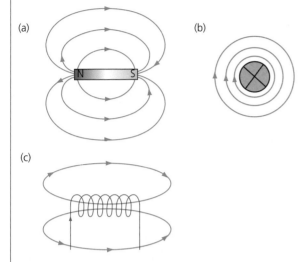

(a)

(b)

(c)

Figure 24.20 (a) The magnetic field around a bar magnet. (b) The magnetic field around a wire carrying current into the paper. (c) The magnetic field around a solenoid.

- The right-hand grip rule, Figure 24.9, shows you the direction of the magnetic field round a wire.
- An electromagnet can be made by putting an iron coil inside a solenoid.
- (H) Fleming's left-hand rule (Figure 24.14) shows the direction of the force on a current carrying wire, placed in a magnetic field: thuMb – Motion; First – Field; seCond – Current. This is the motor effect.
- The force on a wire, length L, carrying current, I, at right angles to a magnetic field of flux density, B, is:

$$F = BIL$$

- The motor effect (as described by Fleming's left-hand rule) is the principle behind the electric motor – a coil in a magnetic field tends to turn.
- The split-ring commutator allows a coil to turn continuously.
- An alternating current causes a loudspeaker coil to vibrate (Figure 24.20).
- When a conducting wire moves through a magnetic field (or the magnetic field changes), a potential difference is induced across its ends.

Exam practice

1 (a) State the unit of magnetic flux density. [1]
  (b) Iron and steel are two types of magnetic material.
    (i) Which material can be used as an induced magnet? [1]
    (ii) Which material can be used as a permanent magnet? [1]
  (c) (i) Sketch the shape of the magnetic field around a bar magnet. [2]
    (ii) Explain how you would use a compass to plot the shape of the magnetic field. [2]
  (d) Figure 24.21 shows two electromagnets A and B, which are suspended on threads and are free to move.

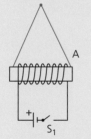

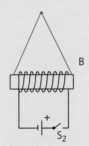

A   B

Figure 24.22

Figure 24.21

    (i) What will happen when switches S_1 and S_2 are closed? Will the magnets attract or repel? [1]
    (ii) What will happen when one of the batteries is reversed, and the switches are closed? [1]

→

2 (a) Copy Figure 24.22 and sketch the shape of the magnetic
field around the wire. [2]

(b) The flux density between the poles of the magnet in
Figure 24.23 (a) is 0.1 T.

   (i) Use the information in Figure 24.23 (a) to calculate
the size of the force on the wire. [2]

   (ii) Determine the direction of the force on the wire. [1]

(c) Explain what will happen to the coil of wire shown in
Figure 24.23 (b), when the current is switched on. [3]

(d) (i) A split-ring commutator is added to the coil of
wire as shown in Figure 24.23 (c). Explain how the split-
ring commutator allows the coil to turn continuously. [2]

   (ii) State two changes to the motor design that
would make the motor turn faster. [2]

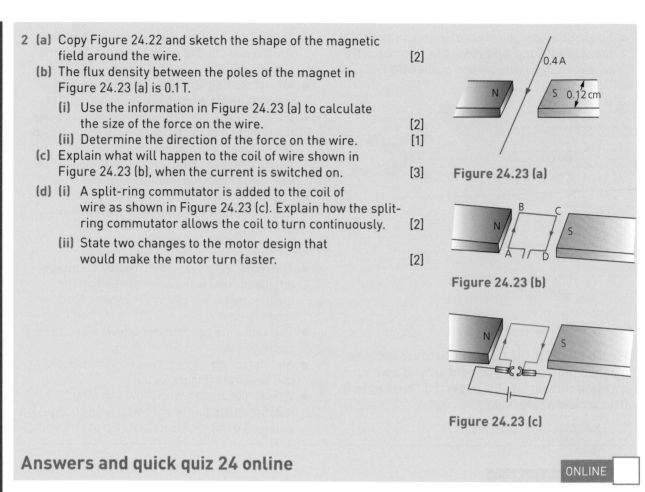

Figure 24.23 (a)

Figure 24.23 (b)

Figure 24.23 (c)

Answers and quick quiz 24 online

ONLINE